51.00

51.00

AN INTRODUCTION TO
NUCLEAR ASTROPHYSICS

GEOPHYSICS AND ASTROPHYSICS MONOGRAPHS

AN INTERNATIONAL SERIES OF FUNDAMENTAL TEXTBOOKS

VOLUME 18

AN INTRODUCTION
TO
NUCLEAR ASTROPHYSICS

The Formation and the Evolution of Matter in the Universe

JEAN AUDOUZE

Institut d'Astrophysique de Paris, France

and

SYLVIE VAUCLAIR

*DAPHE, Observatoire de Meudon, France
and Institut d'Astrophysique, Paris*

D. REIDEL PUBLISHING COMPANY

DORDRECHT : HOLLAND / BOSTON : U.S.A.
LONDON : ENGLAND

Library of Congress Cataloging in Publication Data

Audouze. Jean
    An introduction to nuclear astrophysics.

    (Geophysics and astrophysics monographs; v. 18)
    Enl. and updated translation of L'Astrophysique nucléaire.
    Includes bibliographies and index.
    1. Nuclear astrophysics. I. Vauclair, Sylvie, joint author. II. Title. III. Series.
QB464.A9313 1979   523.01'9'7   79–20752
ISBN 90–277–1012–0
ISBN 90–277–1053–8 (pbk.)

Published by D. Reidel Publishing Company,
P.O. Box 17. Dordrecht, Holland

Sold and distributed in the U.S.A., Canada, and Mexico
by D. Reidel Publishing Company, Inc.
Lincoln Building. 160 Old Derby Street, Hingham,
Mass. 02043, U.S.A.

Printed in The Netherlands

TABLE OF CONTENTS

FOREWORD

This textbook represents an extended version of a short monograph *l'Astrophysique Nucléaire* written by both of us and published by les Presses Universitaires de France in 1972. The text of the present work has been enlarged (the material from the French monograph represents about 33% of this book) and updated because of the impressive progress achieved in this fast moving field. This book, which is also largely based on lectures presented to many students, mainly those of Ecole Polytechnique for Jean Audouze and those of Université Paris VII for Sylvie Vauclair, is intended for undergraduate (astronomy major or non major) and graduate students interested in Astrophysics and also to a more general audience since no specific expertise in astronomy, physics and mathematics is required to fruitfully read this text.

We are both grateful to our many students who have been attending our lectures and forced us to shape the material presented here. We acknowledge the invaluable help of Mmes Annie Dao, Marie-Claude Pantalacci and Madeleine Steinberg in the material presentation of the text. We would like to thank Dr. Billy McCormac, the editor of the collection, and Dr. Gerard Vauclair for their critical comments on the manuscript. Last but not least, we thank the D. Reidel Publishing Company for their patience.

JEAN AUDOUZE AND SYLVIE VAUCLAIR

Paris and Meudon
December 1978

INTRODUCTION

Man has always been intrigued by his environment and by his relationship with it. During the first ages of Humanity this relationship was described or debated by religions or superstitions and not by what we designate under the name of sciences. But, since the very first ages, men have adopted the scientific attitude which is to describe an environment in terms of interactions between the various constituents of what is believed to be the Universe. This is why astronomy, which tries to describe the whole Universe and to understand its evolution, has been the first science to emerge from the human mind. During antiquity men already knew the difference between stars and planets. More recently, the important (and controversial at the time) contributions of Copernicus and Galileo (to choose among the most famous astronomers) are quite well known.

Man has spent more time penetrating the nature of the microcosm than making a rough description of the macrocosm. If the word and the concept of the atom is indeed ancient (atoms were imagined by Greek philosophers such as Democritus), their study in the scientific sense has only been initiated during the 19th century: one must wait for the works of Thompson and Rutherford to get into the tremendous and complex world of the atoms themselves which are composed of almost empty spaces where electrons jingle around extremely small and dense nuclei. At present, many physicists devote their scientific endeavors to try and discover what is the actual basic stone (if any!) of our Universe: the atomic nuclei are constituted by nucleons (protons and neutrons). Current theories on the nature of the nucleons themselves assume that they are formed by a combination of three more elementary or basic particles, the quarks. Research on such particles is presently the subject of very refined and difficult experiments: big accelerators such as those of CERN, Pulkovo, Argonne, ... are daily creating strange particles, but have not yet unravelled their secret.

Nuclear astrophysics attempts to describe how and where matter has been formed. This subset of astrophysics relies heavily on nuclear physics which is a microcosmic description of the matter. In this monograph we attempt to show that the nature and the evolution of macro-objects such as galaxies or stars are directly related to the physical behavior of infinitesimal entities such as atoms and their constituents, electrons, nuclei and nucleons, i.e. protons and neutrons.

One of the best examples of this correlation is the fact that the luminosity of more than 90% of the visible stars (including the Sun) is due to the release of nuclear energy. It is quite obvious in the case of the Sun that other sources of energy are unable to explain its rather old age ($\sim 4.6 \times 10^9$ yr). The discovery of nuclear energy is a direct consequence of the famous Einstein equation which relates mass to energy

($E = mc^2$ where c is the light velocity i.e., 3×10^{10} cm s^{-1}) and of the discovery made by Aston by mass spectrometry techniques that one He nucleus weighs less than four protons. Therefore, when four protons transform into one He nucleus at the center of the Sun, the energy released by such a fusion is sufficient to account for the solar luminosity and it can last for about 10^{10} yr (Chapter IV). Similarly, the explosion of supernovae releases energies as large as 10^{51} erg (i.e. about 10^{10} times the energy released by the Sun during one year. As we will see in Chapter V these explosions might be triggered by mechanisms such as the energy transport by neutrinos. All throughout the monograph there will be a recurrent cycle between the study of the physics of massive objects such as stars ($R \sim 10^{10}$ cm) or galaxies ($R \sim 3 \times 10^{22}$ cm) and the study of nuclear physics which governs the behavior of micro-entities, such as nucleons and nuclei ($R \sim 10^{-13}$ cm). To introduce the reader to this exercise of a double look at very large and very tiny objects, we have gathered in Table I the relative dimensions of some of the different constituents of the observable matter while Figures 1, 2, and 3 respectively show a galaxy (composed of about 10^{11} stars) a small cluster of stars and a view of a 1 μm meteoretical crystal of silicate mineral obtained by electron microprobe techniques. The purpose of this monograph, therefore, is to try and offer a simple and hopefully consistent view on the formation of the different nuclei, which constitute entities as diverse as our bodies, the Earth and all the celestial objects. It will be seen that such studies of the so-called nucleosynthetic processes also suggest answers to various important problems such as the age and the possible origin of the Universe, the evolution of stars and galaxies . . .

The flow of this monograph is as follows: Chapter I describes the observational basis of Nuclear Astrophysics, namely an overview of the observed Universe in relation with the four basic interactions of Physics (gravitation, nuclear, electrical and weak interactions) and the composition of the observed Universe. Chapter II deals with the evolution of matter in the Universe; some features relevant to cosmology are recalled together with a brief review of the stellar and galactic evolution and a quick survey of the main nucleosynthetic processes and the regions of the Universe where these processes take place. Chapter III reviews the abundance of the chemical elements, i.e. the chemical composition of the observed Universe and describes the techniques by which these abundances are determined. The abundances of chemically peculiar stars (Ap and Am stars) are recalled as well as the non nuclear processes (diffusion of the elements) responsible for these anomalies. Chapter IV provides the reader with the principles of nuclear physics which are needed to compute the rate of fusion reactions responsible for the nucleosynthesis of elements from H to Fe. Chapter V analyses the end of stellar evolution, in particular, the mechanisms responsible for the supernovae and novae explosion. The bulk of this chapter is devoted to the so-called explosive nucleosynthesis which takes place during these events and which is able to explain the nucleosynthesis of rare nuclear species such as ^{13}C, ^{15}N, ^{17}O, ^{25}Mg, ^{26}Mg, ^{29}Si, ^{30}Si . . . Chapter VI describes the formation of the elements heavier than Fe which are synthetized by neutron absorption reactions. The main nuclear features of the neutron induced reactions such as the principles of the slow and rapid neutron absorption processes are recalled. A few words are said about the nucleosynthesis of p process elements which are both the rarest and the most proton rich heavy elements. Chapter VII reviews the

TABLE I

Logarithmic scale (cm)

---

30 — The furthest detected radio galaxy

The furthest detected visible galaxy

25 —

Radius of the galaxy: 15 kpc

20 —

Radius of a stellar cluster
distance of the nearest star 1 pc $= 3 \times 10^{18}$ cm
1 light-year $= 10^{18}$cm

15 —

Distance Earth-Sun: 1 astronomical-unit $= 1.5 \times 10^{13}$ cm
Radius of the Sun: 7×10^{10} cm
Distance Earth-Moon: 3.8×10^{10} cm

10 —

Radius of the Earth: 6370 km
Radius of the Moon: 1700 km
Radius of a neutron star: $\simeq 10$ km

5 —

The Man $\simeq 170$ cm

0 —

The Ant $\simeq 0.3$ cm

-5 —

The smallest protozoa 1 μm $= 10^{-4}$ cm
The atomic radius 1 Å $= 10^{-8}$ cm

-10 —

The nuclear radius 1 fm $= 10^{-13}$ cm

-15 —

Fig. 1.   A cluster of galaxies in the region of 73 Leo. The size of this cluster is of the order of 3×10^6 light years. (*Photograph kindly lent by the Observatoire de Haute Provence du Centre National de la Recherche Scientifique, France.*)

Fig. 2a.   A globular cluster of stars: NGC 5272. This cluster of about 10^5 to 10^6 stars has been formed like the 120 other globular clusters in the very early stages of the Galaxy.

Fig. 2b.   An open cluster of stars: h Per (NGC 869–884). This relatively young cluster is inside the Galactic Disk. (*Photographs kindly lent by the Observatoire de Haute Provence du Centre National de la Recherche Scientifique, France.*)

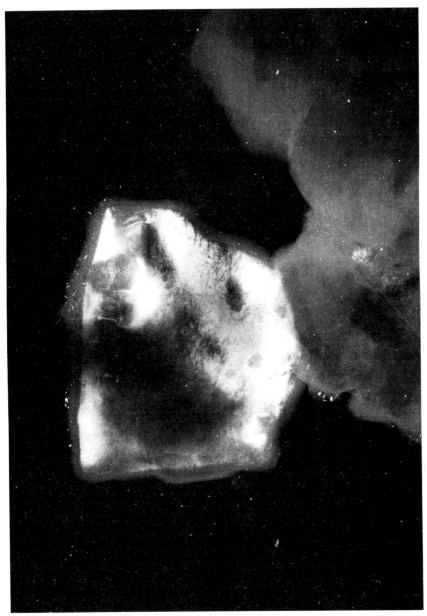

Fig. 3.   Micron size crystal of a lunar feldspar collected in the Luna 16 mission. This crystal has been submitted to the solar particle irradiation: the amorpheous coating and the rounded habit is due to the solar wind, especially the H and He components of about 1 keV (10^3 eV) per nucleon. Inside the crystal damage tracks induced by heavy ion components (especially Fe) of the solar cosmic rays are noticeable. This specific crystal has been submitted to a flux of about 10^{10} heavy ions cm^{-2} (energy of a few MeV per nucleon). This very detailed plate has been obtained and kindly let by the Orsay group (M. Maurette *et al.*) and has been taken with the High Voltage Electron Microscope of the Institut d'Optique Electronique du C.N.R.S. (Toulouse, France).

nucleosynthesis of the lightest elements with atomic masses below 12 (D, ^{3}He, ^{4}He, ^{5}Li, ^{7}Li, ^{9}Be, ^{10}B, and ^{11}B). Some of these elements especially Li, Be and B are formed by spallation reactions induced by rapid (a few MeV) protons or alpha particles impinging on C, N or O nuclei. These reactions which are presented in this chapter occur quite naturally in the interaction between the galactic cosmic rays and the interstellar medium. At present, the hypothesis which is the most successful in explaining the formation of D, ^{3}He, ^{4}He and ^{7}Li is the nucleosynthesis occurring during the primordial phases of the Universe (Big Bang nucleosynthesis). The Big Bang model assumes that the Universe has been very dense and very hot at its birth. Chapter VIII summarizes the methods by which the age of the Universe can be estimated: using the Universe expansion (Hubble age), the position of the stars of the oldest stellar clusters (the globular clusters) in the luminosity-superficial temperature (Hertzsprung-Russell) diagrams, and finally the so-called nucleo-cosmochronological methods based on the search for abundances of long lived radioactive nuclei such as U, Th, Pu and also ^{129}I. Various isotope anomalies have been discovered in some mineralogic phases of carbonaceous chondrites such as the Allende meteorite which fell in Mexico in 1971. These anomalies concern O, Mg, Ca, Nd, Ba ... and can be considered both as big puzzles for all the nucleosynthetic theories and as possible clues to understand the formation of the solar system. The present isotope determinations as well as a few scenarios proposed to try to explain these determinations are given at the end of this chapter. Finally, in Chapter IX the studies of the nucleosynthesis are used to attempt a description of the galactic evolution. The evolution of the density of the stars, of the interstellar gas and of the abundances of the different chemical elements is reviewed as well as the influence of important parameters such as the rate of star formation and the nucleosynthetic power of the stars of a given mass.

At the end of this brief introduction we would like to stress the novelty of the field covered by this monograph. It was only in 1938 that the physicists Bethe and von Weiszäcker began to apply the discovery of the large energy released by nuclear reactions to the physics of the Sun and related stars. The field has indeed become very active after the works of Hoyle in (1946) and Salpeter (1951) which preceded the influential reviews of Burbidge *et al.* (1957) and of Cameron (1957) on the nucleosynthesis occurring in stellar interiors. At about the same time (1949) Gamow tried to push the idea that a large part of the nucleosynthesis occurred during the birth of the Universe in the frame of the Big Bang cosmology. Otherwise, the major developments of nuclear astrophysics are much more recent: for instance, explosive nucleosynthesis has been studied and developed after 1965 (around 1970); the origin of the light elements of atomic mass $A < 12$ has only been understood around 1973; the nucleocosmochronology and the nova explosions, around 1972. Finally, half of the work which deals with the problem of chemical evolution of galaxies has been performed after 1974 according to Audouze and Tinsley (1976). This novelty of the subject is due to the fact that astrophysics is a multidisciplinary field. Its development needs progress in both astronomy and nuclear physics, and more generally, progress in the whole physics such as the behavior of elementary particles like neutrinos, quarks and gluons, and the finest details of hydrodynamics and thermodynamics. In this monograph, we will emphasize the strong influence of almost all the main physics laws on nuclear astrophysics, which attempts to describe the origin of the material Universe.

References

Quoted in the text:

Audouze, J. and Tinsley, B. M.: 1976, *Ann. Rev. Astron. Astrophys.* **14**, 43.
Bethe, H. A.: 1938, *Phys. Rev.* **55**, 103; **55**, 434.
Burbidge, E. M., Burbidge, G. R., Fowler, W. A., and Hoyle, F.: 1957, *Rev. Mod. Phys.* **29**, 547.
Cameron, A. G. W.: 1957, *Publ. Astron. Soc. Pacific* **69**, 201.
Gamow, G.: *Rev. Mod. Phys.* **21**, 367.
Hoyle, F.: 1946, *Monthly Notices Roy. Astron. Soc.* **106**, 343.
Salpeter, E. E.: 1951, *Phys. Rev.* **88** (2), 547.
von Weizsäcker, C. F.: 1938, *Forschungen und Fortschritte* **15**, 159 (in German).
von Weizsäcker, C. F.: 1938, *Weltall* **39**, 218 (in German).

Other general references:

Clayton, D. D.: 1968, *Principles of Stellar Evolution and Nucleosynthesis*, McGraw-Hill Book Co., New York.
Fowler, W. A.: 1967, *Nuclear Astrophysics*, American Philosophical Society, Philadelphia.
Reeves, H.: 1964, *Stellar Evolution and Nucleosynthesis*, Gordon and Breach Science Pubs., Inc., New York.
Trimble, V.: 1975, *Rev. Mod. Phys.* **47**, 877.

ACKNOWLEDGEMENTS

CHAPTER I
Fig. I.6   is from Y. M. Georgelin and Y. P. Georgelin, 1976, *Astron. Astro-phys.* **49**, 74. Permission to reproduce this figure has been kindly granted by the authors and the Editor-in-chief of the journal.

CHAPTER III
Fig. III.6   is Figure 9 from Shapiro and Silberberg, quoted by V. Trimble, Invited Paper of *Proceedings of the Royal Society*, London, 1974. Reproduced with the kind permission of the authors and the publisher.

CHAPTER IV
Fig. IV.1   is Figure 3-1 from *Atomic Nucleus*, by R.D. Evans, McGraw-Hill Book Co., 1955. Used with permission of McGraw-Hill Book Company and the author.
Fig. IV.3   was redrawn by D.D. Clayton from W.A. Fowler and J.J. Vögl, Figure 4-4, p. 298 in *Principles of Stellar Evolution and Nucleo-synthesis*, McGraw-Hill Book Co., 1968. It is reproduced with the kind permission of the authors and the publisher.
Fig. IV.7   was adapted from Figure 1, from R.T. Rood and M.H. Ulrich, 1974, *Nature* **252**, 366. Reproduced with the kind permission of the authors and the publisher.
Fig. IV.9   is Figure 1, p. 123, from an article by G.R. Cauglan in *CNO Isotopes in Astrophysics*, J. Audouze (ed.), D. Reidel Publ. Co., Dordrecht, Holland, 1977. Reproduced with the kind permission of the author.

CHAPTER V
Fig. V.3   is taken from a chapter by L. Rosino in *Supernovae*, D. Schramm (ed.), D. Reidel Publ. Co., Dordrecht, Holland, 1977. Reproduced with the kind permission of the author.
Fig. V.5   is Figure 1-3, p. 7, from *Pulsars*, by R.N. Manchester and J.H. Taylor, Freeman and Co., 1977. Reproduced with the kind permission of the authors and the publisher.
Fig. V.7   is Figure 1, p. 205, from an article by J. Audouze and B. Lazareff in *Novae and Related Stars*, M. Friedjung (ed.), D. Reidel Publ. Co., Dordrecht, Holland, 1977. Reproduced with the kind permission of the authors.

Fig. V.8   is Figure 1 of an article by K.J. Fricke, 1973, *Astrophys. J.* **183**
                  948. Reprinted by courtesy of the author and the Astrophysica
                  Journal, published by the University of Chicago Press. Copy-
                  right 1973, The American Astronomical Society.

Fig. V.9   is Figure 14 of an article by V. Trimble, 1975, *Rev. Mod. Phys.*
                  47, 952. Reproduced with the kind permission of the author and
                  the publisher.

Fig. V.10   is from an article by J. Audouze and B. Lazareff in *Novae and
                  Related Stars*, M. Friedjung (ed.), D. Reidel Publ. Co., Dordrecht,
                  Holland, 1977. Reproduced with the kind permission of the
                  authors.

Figs. V.11a, b   are Figures 2 and 3 from R.C. Pardo, R.G. Couch, and D.W.
                  Arnett, 1974, *Astrophys. J.* **191**, 714. Reprinted by courtesy of
                  the authors and the Astrophysical Journal, published by the Uni-
                  versity of Chicago Press. Copyright 1974, The American Astro-
                  nomical Society.

Fig. V.12   is Figure 1, p. 147 of an article by J.W. Truran in *Supernovae*,
                  D. Schramm (ed.), D. Reidel Publ. Co., Dordrecht, Holland,
                  1977. Reproduced with the kind permission of the author.

CHAPTER VI

Fig. VI.1   is Figure 4, p. 138, of an article by A.G.W. Cameron, 1973, *Space
                  Sci. Rev.* **15**, 121. Reproduced with the kind permission of the
                  author.

Figs. VI.4, 6, and 8 are Figures 7-30, p. 586; 7-20, p. 563; and 7-27, p. 578 from
                  P.A. Seeger, W.A. Fowler, and D.D. Clayton, 1965, *Astrophys.
                  J. Suppl.* **11**, 121. Reprinted by courtesy of the authors and the
                  Astrophysical Journal, published by the University of Chicago
                  Press. Copyright 1965, The American Astronomical Society.

Fig. VI.7   is Figure 7-21, p. 564, from D. Clayton, in *Principles of Evolution
                  and Nucleosynthesis*, McGraw-Hill Book Co., 1978. Reproduced
                  with the kind permission of the author and the publisher.

CHAPTER VII

Figs. VII.1 and 2 are Figures IV-2, p. 54 and II-8, p. 15 from H. Reeves, in *Nuclear
                  Reactions in Stellar Surfaces and Their Relation With Stellar Evolu-
                  tion*, Gordon and Breach, London, 1971. Reproduced with the
                  kind permission of the author and the publisher.

Figs. VIII.5 and 6 are Figures 1 and 6 from J. Audouze and M. Menneguzzi, *Origine
                  des rayonnements cosmiques*, Vol. 4, p. 549, La Recherche, Paris.
                  Reproduced with the kind permission of the authors and the pub-
                  lisher.

Fig. VII.8   is Figure 3 from R.V. Wagoner, 1973, *Astrophys. J.* **179**, 343.
                  Reprinted by courtesy of the author and the Astrophysical Journal,
                  published by the University of Chicago Press. Copyright 1973,
                  The American Astronomical Society.

CHAPTER VIII

Fig. VIII.2   is Figure 1-20, p. 65, from D. Clayton in *Principles of Stellar Evolution and Nucleosynthesis*, McGraw-Hill Book Co., 1968. Reproduced with the kind permission of the author and the publisher.

Fig. VIII.3   is Figure 1 of an article by K. Gopalan and G.W. Wetherhill. 1971, *J. Geophys. Res.* **76**, 8484. Reproduced with the kind permission of the authors and the publisher.

Fig. VIII.5   is figure on p. 75 of an article by D. Schramm. 1974, *Scientific American* **230**, 69. Reproduced with the kind permission of the author and the publisher.

Fig. VIII.6   is from J.H. Reynolds, 1960, *Phys. Rev. Letters* **4**, 8. Reproduced with the kind permission of the author and the publisher.

Fig. VIII.7   is from E.C. Alexander, Jr., R.S. Lewis, J.H. Reynolds, and M.C. Michel, 1971, *Science* **172**, 837. Reproduced with the kind permission of the authors and the publisher.

Fig. VIII.8   is from J.H. Reynolds, 1977, in *Rare Gas Isotopes to Early Solar System History*, Proc. Soviet-American Conf. on Cosmochem. of Moon and Planets, NASA SP 370-2-771. Reproduced with the kind permission of the author.

Fig. VIII.9   is figure on p. 108 of D. Schramm and D. Clayton, 1978, *Scientific American* **239**, 98. Reproduced with the kind permission of the authors and the publisher.

Fig. VIII.11   is Figure 1 of an article by J. Audouze, J.P. Bibring, J.C. Dran, M. Maurette, and R.M. Walker, 1976, *Astrophys. J. Letters*, L185. Reprinted by courtesy of J. Audouze and the Astrophysical Journal, published by the University of Chicago Press. Copyright 1976, The American Astronomical Society.

Fig. VIII.12   is Figure 1, p. 16, of an article by D. Clayton, in *CNO Isotopes in Astrophysics*, J. Audouze (ed.), D. Reidel Publ. Co., Dordrecht, Holland, 1977. Reproduced with the kind permission of the author.

Fig. VIII.13   is Figure 2, p. 393, from D. Schramm, *Supernovae and Formation of the Solar System, in Protostars and Planets*, T. Gehrels (ed.), Univ. of Arizona Press, 1978. Reproduced with the kind permission of the author and the publisher.

Fig. VIII.14   is Figure 1, p. 402 and Figure 1, p. 404, of H. Reeves, *The Big Bang Theory of the Origin of the Solar System in Protostars and Planets*, T. Gehrels (ed.), Univ. of Arizona Press, 1978. Reproduced with the kind permission of the author and the publisher.

CHAPTER IX

Figs. IX.2, 3, 4, and 6 are Figures 2, 3, 4, and 5 from an article by J. Audouze and B. Tinsley, 1976, *Ann. Rev. Astrophys.* **14**. Reproduced with the kind permission of the authors and the publisher.

Fig. IX.5   is Figure 2 from an article by B.E.J. Pagel and B.E. Patchett,

1975, *Monthly Notices Roy. Astron. Soc.* **172**, 13. Reproduced with the kind permission of the authors and the publisher.

Figs. IX.7, 8a, and 8b are Figures 3, 3a, and 3b of an article by L. Vigroux, J. Audouze, and J. Lequeuz, 1976, *Astron. Astrophys.* **52**, 1. Reproduced with the kind permission of the authors and the publisher.

CHAPTER I

THE OBSERVATIONAL BASIS OF
NUCLEAR ASTROPHYSICS

I.1. The Importance of the Four Fundamental Interactions

With the assumption that the same physical laws apply everywhere, the whole Universe may be described in terms of the 'four fundamental interactions' of physics. The first two laws are short range interactions (10^{-13} cm = 1 fermi); the last two laws are long range interactions (α $1/r^2$).

(1) The nuclear interaction or strong interaction rules the behavior of the 'heavy' elementary particles such as nucleons and heavier particles ($m > 1.6 \times 10^{-24}$ g).

(2) The weak interaction governs the physics of particles as electrons, muons and neutrinos, the beta disintegration of the unstable nuclei and the transformation neutron proton.

(3) The electric or photon interaction follows the Coulomb law (α $1/R^2$) where R is the characteristic distance of interacting elements of matter, the basic particle involved in such an interaction being the atom (radius $\sim 10^{-8}$ cm).

(4) The gravitational interaction expresses that material bodies attract each other according to the Newton law which is the algebraic expression of this interaction (also in $1/R^2$). The strength of the interaction is proportional to the masses of the interacting bodies. Therefore, its effect increases when one goes from bodies of relatively small masses such as ours up to galaxies or clusters of galaxies.

The properties of these four basic interactions are summarized in Table I.1. These interactions apply to very specific families of material elements. The nuclear (or

TABLE I.1
The four fundamental interactions of physics

Interaction	Quanta	Potential range cm	Intensity*
Strong	Mesons	$\simeq 10^{-13}$	$\simeq 10$
Weak	Intermediary boson	$< 10^{-16}$	$\simeq 10^{-13}$
Electrical	Photon	$\infty \left(\alpha \frac{1}{r^2} \right)$	$\simeq 1/137$
Gravitational	Graviton?	$\infty \left(\alpha \frac{1}{r^2} \right)$	$\simeq 10^{-38}$

* The intensity of an interaction is a dimensionless quantity used to scale the forces.

TABLE I.2
The elementary particles

		Name of the particle	Particle	Anti-particle	Mass (MeV)	Spin	Boson Fermion
L E P T O N S		Photon	γ	Identical	0	1	
		Neutrino μ	ν_μ	$\bar{\nu}_\mu$	0	1/2	
		Neutrino e	ν_e	$\bar{\nu}_e$	0	1/2	Weak interactions
		Electron	e^-	e^+	0.51	1/2	
		Muon	μ^-	μ^+	105.65	1/2	
H A D R O N S		Mesons:					
		Pion −	π^-	π^+	139.58	0	
		Pion zero	π^0	Identical	134.97	0	Strong interactions
		Kaon +	K^+	K^-	493.7	0	
		Kaon zéro	K^0	K^0	497.9	0	
		Eta	η	Identical	548.8	0	
	B A R Y O N S	Nucléons:					
		Proton	p	$\bar{p}$	938.21	1/2	
		Neutron	n	$\bar{n}$	939.50	1/2	
		Hypérons:					
		Lambda	Λ^0	$\bar{\Lambda}^0$	1115.4	1/2	Strong interactions
		Sigma +	Σ^+	$(\bar{\Sigma}^+)^-$	1189.3	1/2	
		Sigma −	Σ^-	$(\bar{\Sigma}^-)^+$	1197.6	1/2	
		Sigma zéro	Σ^0	$\bar{\Sigma}^0$	1193.2	1/2	
		Xi −	Ξ^-	$(\bar{\Xi}^-)^+$	1321.2	1/2	
		Xi zero	Ξ^0	$\bar{\Xi}^0$	1315	1/2	
		Omega −	Ω^-	$(\bar{\Omega}^-)^+$	1680	?	

strong) interaction applies to a family of particles called the hadrons (lower panel of Table I.2). This family divides itself into two subsets: (1) The mesons which obey the Bose–Einstein statistics (bosons); some of them (the pions) are considered as the 'vector' of this interaction: they are believed to be exchanged (at least virtually) in a nuclear interaction; and (2) The baryons which include the nucleons (mass of $\sim 1.6 \times 10^{-24}$ g or ~ 940 MeV; 1 MeV $= 10^6$ eV; 1 eV $= 1.6 \times 10^{-12}$ erg) and the heavier strange particles called the hyperons; the latter are highly unstable, their lifetime is generally small ($\sim 10^{-6}$ s). A strong interaction is characterized by the fact that its only interacting terms are the hadrons. As will be developed later on, the nuclear interactions are responsible for stellar energy and nucleosynthesis.

The weak interaction necessarily imply leptons. This family is constituted by particles which obey the Fermi–Dirac statistics (fermions) i.e. electron (e^-) and positron (e^+), muons (μ^-, μ^+) and the four known neutrinos (ν_e, $\bar{\nu}_e$, ν_μ, $\bar{\nu}_\mu$). One of the best examples of the weak interaction is the neutron β decay with a characteristic time of ~ 1000 s $(n \to p + e^- + \bar{\nu}_e)$. The 'vector' of the interaction is presumably also a boson (the intermediary boson) the properties of which only begin to be understood. This intermediary boson W seems to exist in three different states W°, W^+ and W^-. Its spin equals 1 and its mass might be very large ($\sim 30\,000$ MeV). This is related to the fact that the range of the weak interaction should be much shorter ($\sim 10^{-16}$ cm) than those of all the other interactions (which would explain its 'weak' character). The mass of the vector is inversely proportional to the range of the interaction according to the Heisenberg uncertainty principle. This interaction is called weak in comparison to the strong one because the relative cross sections in this case are much smaller than the typical nuclear cross sections (see Chapter IV for a definition of the cross section). As an example, the typical neutrino cross sections are $\sim 10^{-44}$ to 10^{-36} cm^2 while the nuclear cross sections are $\sim 10^{-28}$ to 10^{-25} cm^2. This explains why the detection of the neutrinos is so difficult (this point will be developed later on in Chapter III).

The electrical or photon interaction is characterized by emission and/or absorption of a real (but zero mass) particle: the photon. While the range of the two previous interactions is only $\sim 10^{-13}$ cm, the range of an electrical effect is much larger. This interaction overrules the whole atomic physics. As will be seen further on, atomic physics plays a major role in the prediction of the chemical composition of stars and galaxies from their light emission (spectroscopy). Let us also point out its importance in the whole range of chemical reactions: chemistry becomes more and more important in the understanding of major aspects of astronomy. In particular, the interstellar medium is the bulk of important effects such as the formation of various molecules. The characteristics of our more restricted neighborhood: planets, our biosphere and indeed of ourselves, are obviously determined by chemical effects (such as the effect of the acidity in our stomach in relation with our sense of humour!). The electric repulsion between nuclei (the Coulomb barrier effect) is also one of the major effects involved in the fusion nuclear reactions (Chapter IV).

Finally, all the material elements, whatever their mass or their size is, undergo the less efficient but at least always existing gravitational effect. All the mechanical, dynamical, hydrodynamical effects result from this interaction. The shape and appearance of the visible Universe therefore, are the direct consequence of such an interaction. It can be noticed from Table I.1 that its relative strength is extraordinarily weak and this may explain the fact that gravitational waves are very hard to detect. J. Weber, who has devoted much effort to detect them, claimed that he obtained some positive results. However, his interpretation of his measurements has not yet convinced the scientific community.

I.2. A Brief Description of the Observed Universe

The most important observational features of the so-called astronomical objects are summarized here. We leave for Chapter III the important problem of their chemical composition which is the basis of nuclear astrophysics. Evolutionary aspects will be treated in the next chapter.

We live on a planet, the Earth, whose radius is 6400 km and mean density ~ 5 g cm^{-3}. The Earth to Sun distance (1 Astronomical Unit (AU) $= 1.5 \times 10^{13}$ cm or

150×10^6 km) is just adequate for Earth temperatures to be compatible with the maintenance of animated life.

A brief summary of the observational characteristics of the planetary system is given in Table I.3. Two different classes of planets can be distinguished. The telluric planets, i.e., Mercury, Venus, the Earth, and Mars, are the closest to the Sun; they are solid bodies of rather small radius but large density, surrounded by gaseous atmosphere. The outer planets (also called the major planets), i.e., Jupiter, Saturn, Uranus, Neptune, and Pluto, are much farther away from the Sun. They are much larger but less dense. These outer planets are indeed believed to be in a liquid form. Jupiter is known to have an internal source of energy and in this respect is similar to a small star. It is worth noting that Jupiter itself represents 71% of the total mass of the planetary system. If we exclude Pluto which is going to become a satellite of Neptune, all the planets gravitate around the Sun according to the Kepler laws in one plane called the ecliptic plane.

Small bodies, comets and especially meteorites which occasionally fall down on the Earth give extremely rich information on the composition of the solar system and might unravel the still controversial problem of the origin of this system.

This collection of celestial bodies gravitates around our main source of energy, the Sun. The Sun is a very typical star with a mass of 2×10^{33} g (i.e., 750 times more than the whole planetary system and 3.3×10^5 times more than the Earth), and an average radius of 7×10^{10} cm, (i.e. one hundred times more than the Earth). Its external temperature is ~ 5700 K (the Sun is a yellow star) while, from arguments which will be developed later on, its central temperature must be a few. Its mean density is 1.5 g cm^{-3} and ranges from 10^{-14} g cm^{-3} in the external regions (corona, chromosphere and photosphere) up to a few 100 g cm^{-3} in the more internal regions. The solar photons which are seen from the Earth and radiate part of the solar energy towards our planet, are those emitted by the external zones of the Sun. (The radiation is mainly transferred from the central region to the external layers by a succession of photon emission and absorption.) We have, therefore, no direct information on what occurs in the central region of the Sun. As will be discussed later on, the only way to obtain direct information on the central regions of the Sun, is to measure the neutrinos which may be produced there. Inside the Sun, there is a source of energy able to provide a luminosity of 4×10^{33} erg s^{-1} and this since 4.6×10^9 yr. As we will emphasize in Chapter III and have already noted in the introduction, this is one of the major consequences of the nucleosynthesis occurring there. The Sun is like a nuclear plant surrounded by a still hot but inert material (at least in the nuclear sense).

At the solar surface a patchwork of bright irregular granules with a time-varying size of about 1000 km is continuously observed (granulation). Their existence suggests that the external layers (chromosphere and photosphere) undergo convective motions like those observed in boiling water (Figure I.1). The Sun is surrounded by a hot corona (very rarefied gas as hot as 10^6 K) which can be seen during solar eclipses or by using coronographs (Figure I.2). It continuously releases a wind of particles (10^6 particles cm^{-2} s^{-1} with an average velocity of 450 km s^{-1} near the Earth) which fills up the interplanetary medium (also called solar cavity). This interplanetary medium also contains some gas coming from the external interstellar medium and

TABLE I.3
Characteristics of the planets

Names	Symbols	Mean distance to the Sun (AU)	Sidereal orbital period	Sidereal rotation	Mass (M)	Radius (km)	Number of satellites	Density	Remarks
Mercury	☿	0.39	88 days	59 days	0.055	2434	—	5.4	
Venus	♀	0.72	224.7 days	250 days (retrograde)	0.815	6150	—	5.2	
Earth	⊕	1.00	1 yr	$23^h\,56^m$	1.0	6378	1	5.5	Life (for how long?)
Mars	♂	1.52		$24^h\,37^m$	0.107	3380	2	4.0	
Asteroids									
Jupiter	♃	5.20	11.86 yr	$9^h\,50^m$	318	68 700	12	1.3	
Saturn	♄	9.54	29.46 yr	$10^h\,30^m$	95	57 550	10	0.7	} Rings
Uranus	♅	19.2	84 yr	$10^h\,49^m$ (retrograde)	14.5	25 050	5	1.6	
Neptune	Ψ	30.1	164.8 yr	$15^h\,40^m$	17.2	24 700	2	2.3	
Pluto	P	39.5	248.4 yr	6.4 days	<0.18	3000?	?	?	{ Large excentricity and inclination vs. ecliptic plane

1 AU = 1.5×10^{13} cm.
$M_\oplus = 5.977 \times 10^{27}$ g.

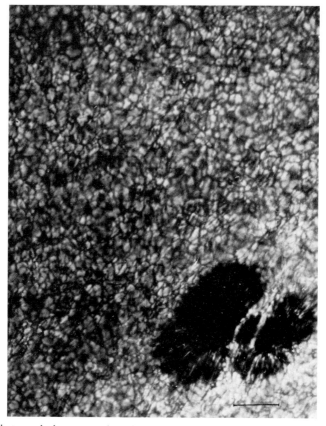

Fig. I.1.   The photograph shows a portion of the solar surface with two solar spots and the patchwork of the solar granulation. The straight line represents about 7000 km on the solar surface. The plate was taken on June 3, 1971, with the Solar Tower of the Meudon Observatory, France, by Mighel, Coutard and Hellier.

some dust concentrated in the ecliptic plane which induces the zodiacal light phenomenon.

The surface of the Sun is where the bulk of transient and sometimes violent phenomena (solar activity) occurs. Dark spots (sunspots) often appear on its surface and can be followed during several solar rotations. They are related to magnetic inhomogeneities. The number of sunspots varies with time and follows an 11 yr cycle.

From time to time, some regions of the solar surface undergo violent events called solar flares which release in a short time (a few hundred seconds) energies from 10^{28} to 10^{30} erg. Solar particle events are associated with some solar flares and from these events originate fluxes of very energetic particles ($>$ a few MeV) (protons and heavier elements) called solar cosmic rays (in opposition with other fluxes of energetic particles which come from outside the solar system and which are called galactic cosmic rays).

Besides the particles, the Sun is also an emitter of radio waves. The reception and

Fig. I.2.   The Solar corona. Photographic observation of the total solar eclipse of September 22, 1968, made by M. Laffineur, M.L. Burnichon and S. Koutchmy, at Yurgamish, Siberia. (*Photograph kindly lent by the Observatoire de Haute Provence du Centre National de la Recherche Scientifique, France.*)

study of such waves help the solar physicist in understanding the rather complex physical processes going on at the solar surface.

The Sun is a slow rotator (it rotates with a period of 28 days). While about all the mass of the solar system is in the Sun, the angular momentum of this system is in the planets.

As has been emphasized before, the Sun is an ordinary star among the about 10^{11} stars which constitute our Galaxy. Table I.4 gives an idea of the existing range in stellar external temperatures, radii and masses. The stars are the astrophysical objects where the bulk of nucleosynthesis takes place. About 90% of them have radii of the same order of magnitude as the solar radius. Some others have radii about 10^2 to 10^3 times larger. They are called red giants because of their low surface temperature (3000 to 4000 K) so that they radiate mainly in the red region of the visible spectrum. A minority of stars, on the other hand, has small radii (comparable to that of the Earth) but their surface temperature can be as high as a few 10^4 K and therefore they radiate in the short wavelength range of the visible spectrum. These stars which have masses similar to that of the Sun are called white dwarfs.

At the end of their evolution stars undergo either steady or violent expulsion of matter. The stars which steadily lose their external layers (or envelopes) are called planetary nebulae (Figure I.3). Another category of stars undergoes a very violent explosion which releases energies as large as 10^{50} to 10^{51} erg during about a month. It constitutes the amazing supernova phenomenon (Figure I.4). A supernova during its maximum has a luminosity which can be 10^9 times that of the Sun. We know from statistical studies that about one supernova every 30 yr explodes in our Galaxy (although only a few of them are observable from Earth). The most famous observed supernovae are: the Crab (explosion in 1054), Tycho (1572) and Kepler (1604). Astronomers would be much interested to see one of such objects exploding in the Galaxy during their own lifetime. After a supernova explosion an object persists which regularly emits pulses (periods between 10^{-3} to a few seconds) in the radio wavelength. These objects are called pulsars and are believed to be very small (up to a few tens km)* but as massive as the Sun (these objects which are identified as neutron stars have densities of 10^{13} to 10^{15} g cm^{-3}, i.e., similar to that of atomic nuclei). The supernova leaves also a supernova remnant which has an expansion velocity of a few 10^4 km s^{-1} (at the beginning of its evolution): this remnant emits synchrotron

TABLE 1.4
Temperatures, masses, radii of some characteristic stars

Star	Stellar category	Temperature (K)	Mass (g)	Radius (cm)
Antares	Supergiant	3 300	3.8×10^{35}	3.7×10^{13}
Arcturus	Giant	4 000	8.4×10^{33}	1.8×10^{12}
η Ori		23 000	2.7×10^{34}	5.0×10^{11}
Sirius A	Main	9 700	6.6×10^{33}	1.7×10^{11}
Sun	sequence	5 800	2.0×10^{33}	6.96×10^{10}
Barnard's star	stars	3 000	7.5×10^{32}	3.5×10^{10}
Sirius B	White dwarf	30 000	1.9×10^{33}	5.0×10^8

* Their radius is deduced from the period of their radiopulses.

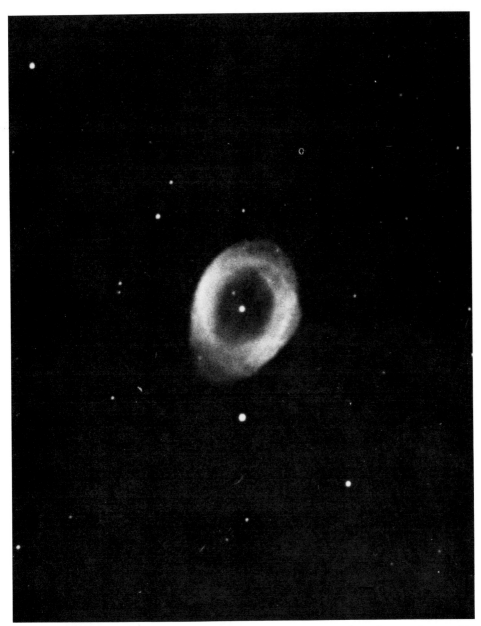

Fig. I.3. A planetary nebula, Lyra (M57, NGC 6720). The nebula comes from the central star and is emitted from it with a velocity of 19 km s^{-1}. (*Photograph taken from the 193 cm telescope of the Observatoire de Haute Provence du Centre National de la Recherche Scientifique, France, and kindly lent by this Observatory.*)

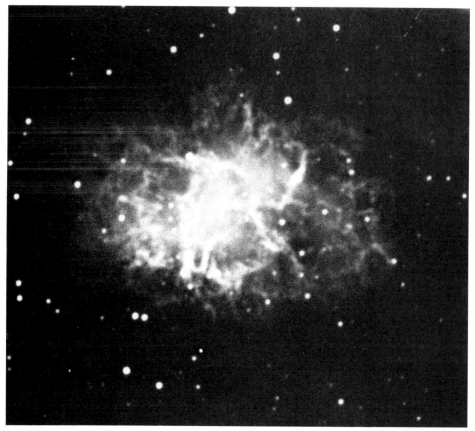

Fig. 1.4.   A supernova remnant; the crab nebula (1054 AD). (*Photograph kindly lent by the Observatoire de Haute Provence du Centre National de la Recherche Scientifique, France.*)

radiation and mixes with the interstellar medium after $\sim 10^5$ yr. Pulsars have the following characteristics:

(A) They emit pulses of radio or visual radiation very regularly (they have been called 'astronomical clocks'). In fact, the period of the pulses slightly increases with time.

(B) Their luminosity is slightly variable with a period several orders of magnitude longer than the period of the pulses.

(C) Their period can drop suddenly and then slowly increase to its initial value (starquakes).

These characteristics seem to be explainable if the pulsars are neutron stars. In particular, the radiation flux could be channeled in one direction by a very intense magnetic field; then the observed pulses would be due to the rotation of the star.

If the star is more massive than about 10 solar masses, it collapses without exploding. Nothing can stop the collapse. When the stellar radius becomes less than a critical value, the 'Schwarzschild radius'* which is computed in the frame of the General Relativity, the radiation itself cannot get out of the star; it becomes a black hole.

Finally, there is another type of exploding objects, more frequently observed ($\simeq 40$ per yr) which undergo less violent outbursts of 10^{43} to 10^{44} erg. They are called novae. These objects belong generally to binary systems (see the next chapter). The last important nova outburst was seen in Cygnus during August 1975. The energy released during this event has been much larger ($\sim 10^{46}$ to 10^{47} erg) than in a typical nova phenomenon.

The majority of stars belong to clusters. The simplest cluster is that constituted by a binary system (or double stars). In this case, the stars are at such a close distance that they gravitate around each other according to the Kepler laws.

The third Kepler law (in the Newtonian form) may be written as:

$$\frac{P^2}{a^3} = \frac{4\pi^2}{G(M_1 + M_2)},$$ (I.1)

where P is the period of rotation, a the semi-major axis, M_1 and M_2 the mass of the two stars. If the first star is much more massive than the second one ($M_1 \gg M_2$), the first star itself can be considered as the center of mass of the system. In this case, the mass of the first star can be calculated from Equation (I.1) (whereas the mass M_2 is neglected).

This method is presently used to try to discover low mass invisible stars around which some stars obviously gravitate. On the other hand, the search for the so-called massive 'black holes' is made by looking at some very energetic binary systems. observed as X-ray sources. This is actually one of the most active and exciting fields of modern astrophysics. Finally, note that the binary systems do not constitute an odd phenomenon: in fact, at least half of the stars belong to a binary system.

Before going further in the definition of stellar clusters, let us describe our own Galaxy. Figures I.5 and I.6 show a schematic view of it. It has the shape of a disk of about 25000 parsecs (1 pc = 3×10^{18} cm) of diameter and 400 pc of thickness which is enlarged at the center. The disk is surrounded by a halo which has a spherical symmetry. The majority of the stars are located inside the disk. They are called population I stars and are generally more blue and younger than the stars of the halo (population II stars). The disk is not homogeneous. Spiral arms can be observed inside the disk. They correspond to a maximum of the star and of the interstellar gas density. These arms rotate around the center with a period of about 10^8 yr.

The so-called galactic clusters or open clusters are found in the disk of our own Galaxy. They are composed of a number of stars ranging from a few tens up to a few thousands: a well known example is the Pleiades cluster which consists of about one hundred stars and has a diameter of a few parsecs. These groups of stars have irregular shapes and do not show any specific concentration at their center. Smaller clusters are called concentrations.

---

* $R_s = 2GM/c^2$ where G is the gravitational constant (6.67×10^{-8} c.g.s.), M is the mass of the body and c is the velocity of the light (3×10^{10} cm s^{-1}).

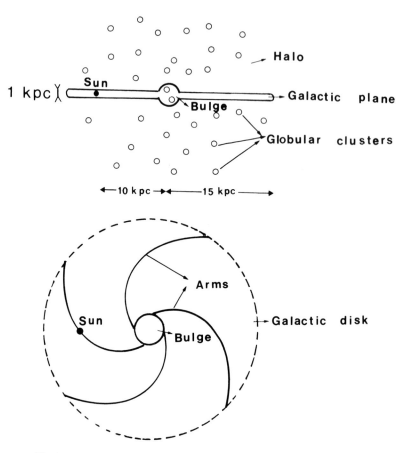

Fig. I.5.   Schematic view of our Galaxy. *Top*: side view, *bottom*: overview.

The second type of clusters is found in the halo. They are the so-called globular clusters which have a somewhat spherical symmetry. They include about 10^4 to 10^5 stars whose density increases towards the center of the cluster. Other distinctions can be made between these two types of clusters. The galactic clusters generally radiate a significant fraction of their light in the blue-yellow range while the globular clusters radiate most of their energy in the red. Furthermore, the globular clusters have very specific dynamical properties. About 120 globular clusters gravitate around the Galaxy and are uniformly distributed around the center of the Galaxy. The discovery of this geometrical characteristic allowed Shapley (1917) to show that the region where the solar system is located is not the center of our Galaxy. The center of the Galaxy is in the direction of the Sagittarius constellation at a distance of 10 000 pc. This discovery of the globular cluster distribution which supersedes the heliocentrism of the Galaxy has been as important in Astronomy as the one made by Galileo which supersedes the geocentrism of the Universe. The clustering of the stars is an important observational fact which means that generally the stars are not borne

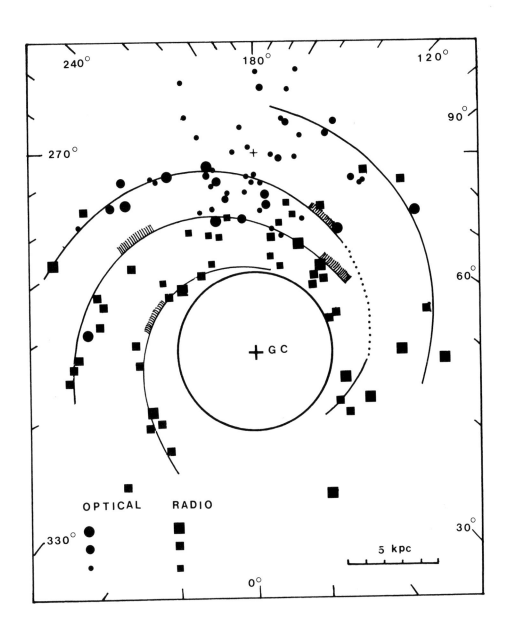

Fig. I.6. The spiral arms of our Galaxy: spiral model of our Galaxy obtained from interstellar ionized hydrogen region; (see further); the resulting spiral pattern has two symmetrical pairs of arms (i.e. four altogether). Major arm: *Sagittarius-Carina*; Intermediate arm: *Scutum-Crux arm*; Internal arm: *Norma Arm*; External arm: *Perseus arm*. Hatched areas correspond to intensity maxima in the radio continuum and in neutral H.

alone: the 'astration' of a given cloud of interstellar matter leads to the formation of many stars at the same time.

In our Galaxy, there is an important amount of matter which is not in the stars and which constitutes the interstellar medium or the interstellar gas. This so-called 'gas' is indeed very tenuous: its density is in average only one particle per cm^3 which corresponds to vacuums that nobody knows how to achieve on Earth. This interstellar gas is not distributed uniformly throughout the Galaxy: it is very patchy and clumpy. The general consensus is to describe it as composed of three different phases. The first one is relatively dense (~ 10 particles cm^{-3}) but cold ($T \leq 100$ K). A typical cloud measures about 10 to 20 pc and has a mass of 1000 solar masses. These clouds are supposed to be regions where new stars can be formed (Orion Figure I.7 is an example of such a region of stellar formation). Between these clouds a very diluted (≤ 0.1 particle cm^{-3}) but hot ($T \geq$ a few 10^3K) intercloud medium exists (second

Fig. I.7. The Orion Nebula. This nebula is located at a distance of about 1500 AL, with a diameter of about 15 AL and a mass of about 300 solar masses. Very young stars are observed in this gaseous nebula which is believed to be the site of formation of new stars. (*Photograph kindly lent by the Observatoire de Haute Provence du Centre National de la Recherche Scientifique, France.*)

phase). The molecular clouds (see below) constitute the third phase. The actual situation is indeed somewhat more complicated, there might be in fact more than three different phases in the interstellar medium.

There is more gas in the solar neighborhood than in the galactic center. The relative fraction of gas represents $\sim 10\%$ of the total mass in regions such as the solar neighborhood while it represents only less than $1/100$ in the galactic center.

The interstellar gas cannot be considered as very 'clean'. There are not only many different atoms or ions (the ions can be found for instance in the vicinity of hot stars whose radiation is able to ionize a few atoms such as H, C or He) but there are many different molecules (up to forty five now) which have been discovered in the interstellar medium (Table I.5 and Figure I.7). Some are rather simple such as OH, H_2O, CO, or NH_3, but many are rather complex organic molecules (HC_7N). Even the ethylic alcohol which has some well known effects on earth has been evidenced in the interstellar medium. Moreover, the interstellar gas is dusty. A significant fraction (1 to 10%) of the interstellar gas is constituted of small solid grains which are called 'dust' by the astronomers. In Figure I.7, one can see the presence of dust in Orion. This dust component is still poorly understood but considerable work is being undertaken to understand its composition and the effect of this solid component on the physics of the interstellar medium. Presently, it is believed that these grains are small (some fractions of a μm) and have a mass of about 10^{-13} g. Their composition is presumably some silicated ice with some trace of heavier elements such as Fe, surrounded by envelopes containing mainly CNO atoms.

TABLE I.5

Known interstellar molecules as of January 1977
(This list is increasing every month by more than one molecule.)

Diatomic (2 atoms)	Triatomic (3 atoms)	4 atoms
H_2, OH, SiO, SO, SiS, NS, CH^+, CH, CN, CO, CS,	H_2O, H_2S, SO_2, HCN, OCS, HCO^+, HCO, CCH, NH_2^+	NH_3, H_2CO, HNCO, H_2CS, C_3N

5 atoms	6 atoms	7 atoms
HC_3N, HCOOH, CH_2NH, H_2CCO, NH_2CN	CH_3OH, CH_3CN, NH_2CHO	CH_3C_2H, CH_3CHO, NH_2CH_3, CH_2CHCN, HC_5N

8 atoms	9 atoms	
$HCOOCH_3$	$(CH_3)_2O$, $CH_3CH_2OH^*$, CH_3CH_2CN	

* Ethyl alcohol.

The interstellar medium produces copiously and is traversed by many different radiations:

(1) radio-emission (we will come back to the very well known 2.8 K radiation); 21 cm line (coming from hydrogenic clouds) and molecular lines (18 cm line of OH);

(2) IR radiation (coming from molecular spectra or reemitted by the dust);

(3) UV produced by hot stars; and

(4) X-ray or gamma-ray emission coming from hot spots of the Galaxy such as the supernova remnants.

Energetic particles (from a few MeV up to 10^{20} eV) which are called galactic cosmic rays and which are described later (Chapter VII) irradiate continuously and homogeneously the interstellar gas (where they induce spallation reactions) before arriving into the solar system. The interstellar medium contains magnetic fields of 10^{-5} to 10^{-6} G and is traversed by hydromagnetic waves which govern the physics of this medium.

The interstellar medium, although its density is very low, appears to be a very complex medium. A tremendous effort is made by astronomers and astrophysicists to understand this complexity. This shows how many different fields of physics have to be used in astrophysics to understand the Universe: atomic physics, solid state physics, plasma physics, inorganic and organic chemistry, all of them are important in the study of such a complex medium.

Now let us describe these large concentrations of matter which are the galaxies. A very crude estimate of the number of possibly observed galaxies is about 3 to 10×10^9. They can be classified like animals or plants (Figure I.8). Such a classification has been made by the American astronomer Hubble (Table I.6) who initiated the construction of the well known 5 m diameter telescope of Mount Palomar and made many important discoveries about galaxies. Galaxies can be roughly divided into three different classes, each of them including several sub-types.

First, the elliptical galaxies (Figure I.9a) have a spherical or elliptical shape and are basically constituted of stars (there is almost no gas inside them) which radiate more in the red than in the blue. These stars seem to be homogeneously gathered inside such galaxies, although their concentration increases from the peripherical

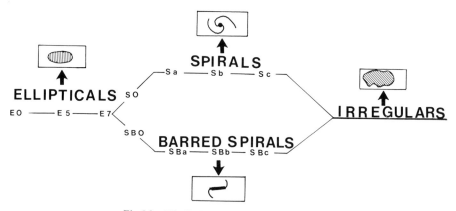

Fig. I.8. The Hubble classification of galaxies.

TABLE I.6

Some characteristics of the different types of galaxies in the Hubble classification. M/L represents the mass-luminosity ratio, while M_g/M represents the fraction of the total mass in the gaseous form.

Type	Spectral type of the central region	M/L	M_g/M
E	G4	80	$\lesssim 10^{-3}$
So	G3	50	
Sa	G2	30	
Sb	G0	20	0.05–0.1
Sc	F6	10	0.15
Irr	—	3	0.3

regions to the central regions. Such objects are relatively massive ($\geq 10^{12}\ M_\odot$), (there are however exceptions like dwarf ellipticals). In contrast with other classes of galaxies, they radiate less light compared to their mass (Table I.6).

The second class (70% of them) is that of spiral galaxies (Figure I.9b). Our own Galaxy is typical of this group. Spiral galaxies have a mass of the order of $10^{11}\ M_\odot$ and a relative ratio gas/total mass of the Galaxy of ~ 0.1. These galaxies are relatively more luminous and bluer than the elliptical galaxies (in comparison with their masses). Among the spiral galaxies there is a large variety of shapes, some of them showing a bright bar from which the arms originate (sub-class of the barred galaxies*) (Figure I.9c). There exists also a class of rather ambiguous objects, the S0 type galaxies, whose shape is similar to that of the spiral galaxies but which do not show any spiral structure. These objects are still rather difficult to classify. The center of the spiral galaxies, including ours, is similar to the elliptical galaxies themselves (red population, spherical structure and gas).

The third class contains the objects with a not well defined shape (Figure I.9d). These objects are relatively less massive than the other galaxies ($\sim 10^{10} M_\odot$). They are more luminous, bluer and seem to have relatively more gas than the others.

Like the stars, the galaxies seem to be distributed according to some sort of clustering: a large number of galaxies have companions or satellites. Our Galaxy has two satellites, the Large Magellanic Cloud and the Small Magellanic Cloud, which are classified as irregular galaxies. This clustering of galaxies is now well established.

The galaxies carry and release enormous amounts of energies. A fair fraction of galaxies is known to be large radio source emitters. Some of them are more powerful radio sources than the others (this is the case for many elliptical galaxies which have a radio luminosity of $\sim 10^{47}$ erg s^{-1}). There is a specific class of active galaxies called the Seyfert galaxies whose nucleus is especially active.

As we shall see in the next chapter, another important parameter is the relative

---

* The interested reader is referred to the Hubble atlas which exhibits instructive pictures of different types of galaxies.

Fig. I.9 (b).

Fig. I.9 (a).

Fig. I.9a–d. (a) An elliptical galaxy (type E0), (NGC 4486). (b) A spiral galaxy (NGC 628 M74). (c) A barred spiral galaxy (NGC 5905). (d) An irregular galaxy (NGC 4631). (*Photographs kindly lent by the Observatoire de Haute Provence du Centre National de la Recherche Scientifique, France.*)

Fig. I.9 (c).

Fig. I.9 (d).

redshift of the galaxies Z^* ($\Delta\lambda/\lambda \simeq Z$) which characterizes their velocity respective to us. The redshifts of galaxies range from 0 up to $\simeq 1$.

A new class of objects has been recently discovered (~ 1963). They have the same optical properties as the stars. They are not resolved and they show very large redshifts. This is why they are called quasi-stellar objects (QSO's or following the usual wording, quasars). These objects are presumably either a specific type of very active galaxies (or nuclei of galaxies) or galaxies in an early stage of their evolution.

To end this rather sketchy description of our observable universe, it is important to realize (and that may be the main purpose of this book) that the evolution of the whole universe is governed by effects acting on particles as small as nuclei or elementary particles.

References

Quoted in the text:

Shapley, H.: 1917, 'Studies Based on the Colors and Magnitudes in Stellar Clusters', different papers published in *Astrophys. J.* **45** up to **54**, from 1917 to 1920.

Further readings:

Abell, G.: 1964, *Exploration of the Universe*, Holt, Rinehart and Winston Inc., New York.
Allen, C. W.: 1973, *Astrophysical Quantities*, 3rd edition, The Athlone Press, London.
Balian, R., Encrenaz, P., and Lequeux, J.: 1975, *Atomic and Molecular Physics and the Interstellar Matter*, North-Holland Publ. Co., Amsterdam.
Blatt, J. M. and Weisskopf, V. F.: 1952, *Theoretical Nuclear Physics*, John Wiley & Sons, Inc., New York and London.
Encrenaz, P.: 1974, *Les Molécules Interstellaires*, Neuchâtel: Delachaux et Niestlé, in French.
Evans, R.D.: 1955, *The Atomic Nucleus*, McGraw-Hill Book Co., New York.
Field, G. B. and Cameron, A. G. W.: 1975, *The Dusty Universe*, Neale Watson, Academic Pubns., Inc., New York.
Kleczek, J.: 1976, *The Universe*, D. Reidel Publ. Co., Dordrecht, Holland.
Lang, K. R.: 1974, *Astrophysical Formulae*, Springer-Verlag, Heidelberg.
Pasachoff, J. M. and Kutnev, M. L.: 1978, *University Astronomy*, W. B. Sanders Co., Philadelphia.
Sciama, D. W.: 1975, *Modern Cosmology*, The Cambridge University Press, Cambridge.

See also the special editions from Scientific American: *Frontiers in Astronomy, New Frontiers in Astronomy* and *The Solar System*, W. H. Freeman and Co., San Francisco.

---

* If an object which radiates with a wavelength λ moves with respect to the observer with a velocity v, the wavelength of the radiation is shifted by $\Delta\lambda$

$$\frac{\Delta\lambda}{\lambda} = \left(\frac{c+v}{c-v}\right)^{1/2} - 1, \quad \text{or} \quad \frac{\Delta\lambda}{\lambda} = \frac{v}{c} \quad \text{if} \quad v \ll c,$$

where c is the light velocity.

CHAPTER II

THE EVOLUTION OF MATTER IN THE UNIVERSE

II.1. The Origin of the Universe

It is now generally believed that our Universe originated some 10^{10} yr ago in a singular event which is referred to as the 'Big Bang'. This theory is based on the following fundamental observations:

(a) The spectral analysis of the radiation emitted by galaxies shows that all the absorption lines are redshifted (shifted to longer wavelengths, or lower frequencies). Interpreted as a Doppler effect, this observation suggests that the galaxies are moving away from us. Hubble (1929) found that the redshifts of galaxies (hence their velocities) are proportional to their distances: the further the galaxies, the larger their redshifts and therefore their velocities). The relation between the relative velocity of a galaxy and its distance can be written simply as:

$$V = HD,$$

where V is the velocity and D the distance. The so-called 'Hubble constant' H is presently determined as $\simeq 50$ to 100 km s^{-1} Mpc^{-1}. In fact, this 'constant' decreases with time in any cosmological model (except the now ruled out steady state model).

The Hubble law has been interpreted in cosmology as a consequence of the expansion of the Universe. To understand this effect let us make an analogy. If one draws some points at the surface of a balloon which is blown up, the points will now move away one from another during the blowing up. The motion of these points is characterized by a relation similar to the Hubble law: galaxies behave in the Universe like these points. The difference is that the galaxies move in a three dimensional space while the points move in a two dimensional one (the balloon surface).* Now if the time is reversed, one can logically think that the Universe was once concentrated in a single dense spot: the primeval fireball.

(b) The second fundamental observation is due to Penzias and Wilson (1965)** For their observations they used a horn-shaped antenna, first designed to receive signals deflected by Echo satellites, and then modified for radio astronomy. When they happened to detect an isotropic unknown signal, they first attributed it to the noise coming from the antenna and tried to get rid of it. But further on, they realized that the signal was due to extraterrestrial radiation. They measured this signal at

---

* In fact, as it will be seen later on, the analogy with a balloon corresponds to a closed Universe. For an open Universe, the analogy should be made with an infinite surface like for instance a horse saddle shape.
** Nobel Prize winners in 1978.

different wavelengths and deduced from these observations that this radiation was the same as the one emitted by a blackbody with a temperature of 2.8 K (Figure II.1). This is in accordance with what is predicted from the computation of the radiation left by the Universal fireball.

(c) The mere observation that during the night the sky is dark is referred to as the Olbers paradox: in an infinite and static Universe the night sky should be bright due to the luminosity of an infinite number of stars. This paradox can be understood in three different ways: (1) a static but optically thin Universe; (2) a lot of absorbing material between the stars and us; and (3) an infinite but expanding Universe (in this case, above a certain distance the redshifted luminosity is visible no more). Only case (3) agrees with further observations.

(d) As it will be seen later on, there are at least two elements, He and D, which cannot be synthetized with their observed abundances either in stars or in inter-stellar matter. The easiest way to produce these elements in amounts compatible with the observations is precisely the nucleosynthesis occurring during the pri-mordial explosion (Chapter VII).

The most widely accepted model proposed for the birth of the Universe is due to Gamow and Lemaître (1949). It accounts for the redshifts of galaxies and for the 2.8 K universal radiation. In this model, the Universe is extremely hot and dense ($T > 10^{13}$ K and $\rho \sim 10^{15}$ g cm^{-3}) at the origin of times (the initial explosion). Then it cools down and expands, a phenomenon which is still going on nowadays. The Universe is assumed to be homogeneous and isotropic, which is not in con-tradiction with the observations of galaxies. The evolution of the Universe may be divided roughly into four periods (Table II.1) (Figure II.2).

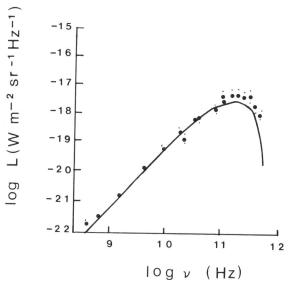

Fig.II.1. The black body radiation at 2.8 K. The abscissa represents the logarithm of the frequency. The ordinate represents the logarithm of the brightness of the radiation. The points represent the observed radiation.

TABLE II.1
The four periods of the evolution of the Universe

Time	Density	Temperature	Energy per particle
Chaos			
Hadron Era			
10^{-4} s	10^{14} g cm^{-3}	10^{12} K	100 MeV
Lepton Era			
10 s	10^4 g cm^{-3}	10^{10} K	1 MeV
Radiation Era			
10^6 yr	10^{-21} g cm^{-3}	3×10^3 K	0.3 eV
Stellar Era			
10^{10} yr	10^{-30} g cm^{-3}	3 K	3×10^{-4} eV

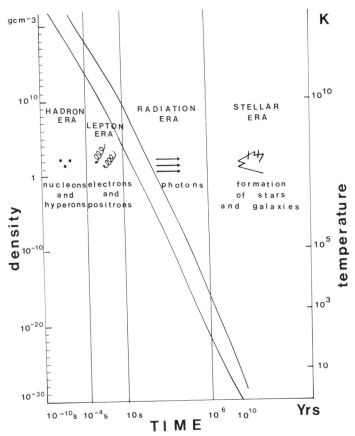

Fig.II.2. Variations with time of the temperature and the density at the origin of the Universe.

II.1.1. THE HADRON ERA

At the very beginning of the Universe, the temperature and the density were very high. The cosmic matter was extremely complex and consisted of particles with energies in the range of 10^{12} eV which is an entirely unknown energy range. The temperature of a radiation evaluated in degrees Kelvin can be related to the energy of the radiating medium in eV (1 eV $\simeq$ 11 000 K) and can be compared to the mass scale of the different elementary particles (Table I.2). Particles may indeed appear in the radiation field when the thermal energy of the medium is higher than their mass energy. Thus, when the temperature of the medium was $\geq 10^{13}$ K (or 1 GeV), the radiation was composed of hadrons (nucleons, mesons, etc. ...), leptons (electrons, muons) and photons. This first period, called the Hadron Era, ends when the thermal energy becomes less than the mass energy of the lightest hadron the π-meson, i.e. about 100 MeV. This happens after 10^{-4} s. At this time, the density of the Universe is of 10^{14} g cm^{-3} and its temperature is 10^{12} K.

Another important parameter concerning the evolution of the Universe is the so-called baryon number, i.e. the difference between the number of baryons (matter) and the number of antibaryons (antimatter):

$$\Delta B = B^+ - B^-.$$

This number is constant with time. At the end of the Hadron Era, baryons and antibaryons annihilate so that the number of baryons which are preserved are at least equal to the baryon number ΔB*.

II.1.2. THE LEPTON ERA

When the temperature becomes less than 10^{12} K (100 MeV), the number of hadrons decreases considerably so that the bulk of the radiation is composed of leptons (electrons, muons and neutrinos) and photons. The basic physical effect which happens at the end of this period is the neutrino decoupling: the characteristic time of interaction of the neutrinos with the matter decreases as the Universe expands. When this time becomes longer than the age of the Universe, the neutrinos do not interact anymore with matter. Cosmological models suggest that the present Universe should be imbedded in a neutrino radiation of about 2 K, residual of this period. The Lepton Era ends after about 10 s, when the thermal energy becomes less than about 0.5 MeV which is the mass of the electron i.e. the lightest lepton). The density of the Universe is at that time 10^4 g cm^{-3}.

II.1.3. THE RADIATION ERA

When the temperature is less than 0.5 MeV, the number of leptons strongly decreases while the photons become predominant. At the end of this radiation era photon decoupling occurs: when the characteristic time of interaction between

---

* Readers interested in the influence of the baryon number on the evolution of the Universe should consult Omnes and Puget (1974).

photons and matter becomes longer than the age of the Universe, the photons cease to interact with matter (in general). They just stay there and are responsible for the 2.8 K black body radiation which was first detected by Penzias and Wilson (1965).

At the end of the radiative era (after 10^6 yr) the temperature of the Universe is about 3000 K and the density 10^{-21} g cm^{-3}.

It is during the radiative era that ^{4}He and ^{3}He have been formed, as well as D, and probably ^{7}Li (Chapter VII), while the majority of heavy elements are synthetized later in stars. (Helium is also synthetized in stars, but in too small a quantity to account for the observed universal abundance: as stated before, the observed He was mainly formed at the origin of the Universe.)

II.1.4. THE STELLAR ERA

After the radiation period the stellar era begins. It is still going on to-day (about 10^{10} yr). The matter left after the three prestellar periods has evolved into galaxies and stars, in a process which is still not very well understood. The present density of the Universe is about 10^{-30} g cm^{-3} and the temperature is of 2.8 K. As expressed by Lemaître (1949), "Standing on a cooled cinder, we see the slow fading of the suns, and we try to recall the vanished brilliance of the origin of the worlds."*

The discovery made by Hubble that the redshift of a galaxy (and therefore its velocity) increases with the distance, is a strong argument in favor of the present expansion of the Universe. But this expansion motion is decelerating with time because of the presence of the matter itself which prevents a free expansion by self-attraction effect due to gravity.

The simplest cosmological models such as those of McCrea and Milne (1934)** show that the presently expanding Universe can be either expanding forever (open Universe as shown in Figure II.3) or experiencing a possible succession of expansion and contraction motions (closed Universe Figure II.3). The deceleration effects are increased with the amount of matter present in the Universe which means that denser universes are closed whereas more tenuous ones are open. One can therefore recall the law ruling the overall motion of the Universe to its present density (which scales the density of the Universe at any given time). Current calculations performed with this model show that if the present density of the Universe is lower than $\rho_0 \simeq 2 \times 10^{-29}$ g cm^{-3} (called the critical density) the Universe is open while it is closed if the actual density of the Universe is larger. In fact, the density deduced simply from galaxy observations is 10^{-31} g cm^{-3}. The question is therefore to know if there is in the Universe a large quantity of 'missing mass' in a nonobservable form such as black holes or big invisible planetesimal bodies which put the actual density of the Universe up to values larger than the critical value. It will be seen in more detail in

---

* The translation is due to Harrison (1968).

** In their models they assume that the dynamics of the Universe can be described in the frame of the Newtonian Universe: infinite, homogeneous and isotropic. The results found in this model are qualitatively similar to those obtained in more correct cosmological models studied in the frame of the General Relativity

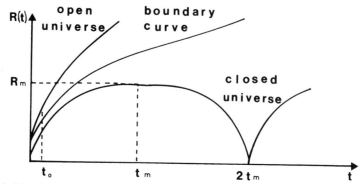

Fig.II.3. Models of open and closed universes. The radius of the curvature of the Universe is plotted vs time. For an open universe the radius increases indefinitely. For a closed universe the radius increases until it reaches a maximum R_m at time t_m and then decreases towards zero at $2t_m$. It can then increase again if the Universe is pulsating. From the evidence now available, it seems that we live in an open universe.

Chapter VII that the presence of D in the observable interstellar medium in abundances as large as $D/H \simeq 10^{-5}$ allows one to conclude that the observable Universe is open (if one believes as in current nucleosynthetic theories that D was produced as ^{3}He during the Big Bang).

II.2. Stellar Evolution

II.2.1. THE HERTZSPRUNG-RUSSELL DIAGRAM

In antiquity, the stars were classified by a scale of magnitudes according to their brightness: the first magnitude corresponding to the brightest stars, and the sixth to the faintest eye-observed stars. Modern astronomers have made this scale quantitative. The measurable quantity is the radiative flux received per surface unit and per time unit on the Earth (this quantity is called the 'brightness' and is labeled E). Due to the response of the eye (logarithmic in apparent brightness), the magnitude scale can be approximated by a logarithmic law in brightness. A difference of 5 in magnitude corresponds to a factor of about 100 in brightness, therefore the difference in magnitude m_2 and m_1 between two stars of brightness E_2 and E_1 is given by the Pogson law:

$$m_2 - m_1 = -2.5 \log (E_2/E_1). \tag{II.1}$$

Note that the brighter the star, the smaller its magnitude. An absolute scale of 'apparent magnitude' is defined by choosing the magnitude of a group of well known stars around the north pole (do not confuse this 'absolute scale' with the notion of 'absolute magnitude' which is described in the sequel). The magnitude defined this way (or 'apparent magnitude') depends on the distance of the star. Astronomers have defined another magnitude scale which does not depend on the stellar distance: the 'absolute magnitude' of a star is the magnitude it would have if it were located at 10 pc away from the Sun. The relation between the absolute and apparent magnitudes is given by:

$$M - m = 5 - 5 \log d,$$ (II.2)

where M is the absolute magnitude, m the apparent magnitude and d the distance.

The measured brightness E depends in fact on the sensitivity of the receiver, which generally varies with the wavelength of the radiation. The only experimental way to get rid of that effect is to use a receiver called bolometer, which measures the total flux received from the star (for the whole wavelength range). The magnitude derived this way is called 'bolometric magnitude'. This can be done for very bright stars. For other stars, the 'bolometric magnitude' is derived by making theoretical corrections to some optical magnitude. This magnitude is directly related to the stellar luminosity (energy spent by the star per second):

$$M_2 \, \text{bol} - M_1 \, \text{bol} = -2.5 \log \frac{L_2}{L_1}.$$ (II.3)

Another fact we can derive from the observation of stellar radiation is its color, which is directly related to the external temperature: red stars are colder than blue ones. A quantitative measurement of the stellar color is obtained by the use of filters. A classical scale is the 'UBV' system, which uses 3 filters:

> U centered at 3500 Å (UV)
> B centered at 4300 Å (blue)
> V centered at 5500 Å (yellow or visible).

The 'color' is expressed by the differences in magnitude $U-B$ and $B-V$ (or 'color indexes').

The stars are also classified according to the lines observed in their spectra ('spectral classification'). The intensity of these lines depends on the stellar external temperature. The stars are traditionally designed by their 'spectral types', which are, from the hotter to the cooler stars:

> O, B, A, F, G, K, M,

(see Figure II.4).

Hertzsprung in 1911 and, independently, Russell in 1913, had the clever idea to plot the stars in a diagram with the external temperature in abscissae (increasing to the left) and the luminosity in ordinates (increasing upwards).

This well known diagram, now called the HR diagram from the initials of its authors, constitutes one of the most powerful tools of modern astrophysics. Using such a diagram if one plots a random sample of stars, one finds that about 80% of these stars gather on a diagonal called the 'main sequence' (Figure II.5). Some others, on the top right, luminous and 'cold' (3000 to 4000 K) are the red giants, while the few hot ones ($T > 10^4$ K) not very luminous, down left, are the white dwarfs. From the well known 'blackbody' relation, one can easily deduce that the stellar radius increases from the lower left towards the top right, according to the relation:

$$L = 4\pi R^2 \sigma T_e^4$$

where L is the luminosity, T_e the 'effective temperature' (approximately equal to the surface temperature), R the radius of the star, and σ the Stefan constant, $\sigma =$

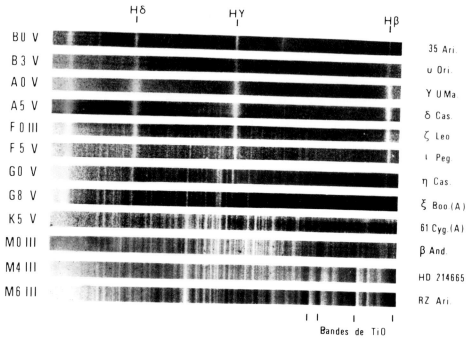

Fig. II.4.   The corresponding surface temperatures are for B0 spectral type, 28 000 K; B3, 13 000 K; A0, 9900 K; A5, 8500 K; F0, 7000 K; F5, 6600 K; G0, 6000 K; G8, 5300 K; K5, 4100 K; M0, 3200 K; M4, 2800 K; M6, 2400 K. Class V corresponds to main sequence stars while Class III corresponds to giant stars. (*Spectra taken with the small objective prism of 16 cm of the Observatoire de Haute Provence du Centre National de la Recherche Scientifique and kindly lent by this observatory.*)

5.67×10^{-5} c.g.s. The experimental HR diagram is well accounted for by the theory of stellar evolution. Stars evolve by varying their luminosities and external temperatures. The region where stars statistically concentrate in this diagram is the one where they evolve very slowly: this constitutes the main sequence which is composed of stars stabilized by the H nuclear burning.

II.2.2. STELLAR EVOLUTION

Stars are formed out of the galactic gas (or 'interstellar matter'). The stellar formation is believed to take place preferably in the galactic arms (cf. Chapter I), where the interstellar matter is compressed. However, evidence of stellar formation is also found in a patchy distribution all through the galactic disk. If the radius of a spherical mass of gas is forced to decrease under a critical value computed by Jeans, it becomes unstable, collapses, and leads to a protostar. At the beginning, the acceleration of the particles varies like the inverse of the square of the density (free-fall). When the radius of the protostar is not too large ($R \simeq 10^5 R_\odot$), the density is $\sim 10^{-15}$ g cm^{-3}, the collapse is rather rapid (~ 500 yr) and the 'protostar' becomes very luminous. Then, the pressure forces inside the protostar begin to brake the gravitational

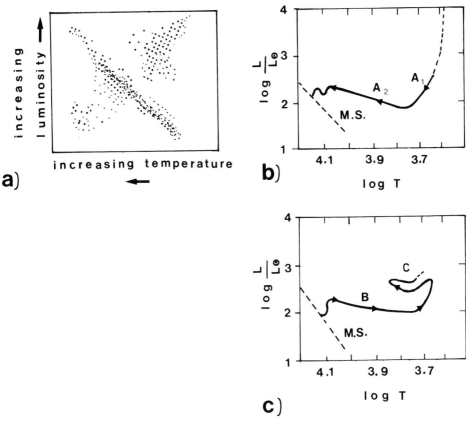

Fig.II.5. The Hertzsprung-Russell diagram. (a) The observed HR diagram for a random sample of stars. (b) The evolution of a $5M_{\odot}$ star towards the main sequence. (c) The evolution of the same star after the main sequence. A $5M_{\odot}$ star stays approximately 10^8 yr on the main sequence (dashed line).

collapse. The star goes on contracting in a thermostatic time (of the order of 10^7 yr). Its luminosity decreases while its external temperature remains approximately constant. The star is then entirely convective (like a boiling magma) and it may undergo violent events: the so-called 'T-Tauri stars' are presently in this phase of their evolution, as well as the 'flare stars', those very strange stars, for which the luminosity increases by a factor of 100 during a few seconds! After this active period the star becomes more quiet. It stops 'boiling' in its center. Then, the energy is transported by the radiation from the center of the star to its surface (the star is in a radiative phase). In the HR diagram, there is a turn off of the stellar track when the star becomes radiative (Figure II.5). The luminosity no longer decreases. On the contrary, it increases very slowly because the increase effect due to the rise of the external temperature is slightly more important than the decrease effect due to the contraction of the star. This evolution of young stars in the HR diagram is called the 'Hayashi track', for C. Hayashi who computed it first. Then, the temperature and the density at the center of the star become high enough for H nuclear burning to take place (some millions of degrees and some grams per cubic centimeters). The condition of stellar equilibrium is that the energy gains equal the energy losses:

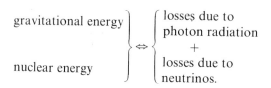

Thus stellar evolution is divided into different alternative periods during which gravitational or nuclear energy are in turn the basic energy source. During the H nuclear burning (which lasts 10^{10} yr for the Sun, less for a more massive star*) the representative point on the HR diagram moves very slowly. This very slow motion explains the dispersion in the observed stars of the points figuring the main sequence, which is then observed as a band and not as a line (Figure II.5). When about 10% of the H has been transformed into He, the nuclear energy is no more sufficient to compensate the energy losses. There is first a short phase of overall contraction, during which the temperature increases inside the star. When the temperature is high enough for H to burn outside the stellar core ('shell burning'), the outer zones expand while the core goes on contracting. The stellar radius increases, the external temperature decreases due to the expansion, and the luminosity decreases very slowly according to the mass-luminosity relation (II.4) – due to opacity effects – until the star becomes highly convective again:

$$L = \frac{\mu^4 m^3}{k},$$

(II.4)

where μ is the mean molecular weight of the stellar gas (μ is the ratio of a given mass of gas by the total number of particles (atoms, ions, and electrons ...) which are in it), and k its mean absorption coefficient per gram. Then, the luminosity increases quickly, the external temperature stabilizes, and the track goes on more or less parallel to the Hayashi track. The representative point in the HR diagram stops when the central temperature becomes high enough for He to burn into C and O (temperature of ~2×10^8 K). The star is then a *Red Giant*. The He nuclear burning is quicker than the H burning; it only lasts about a few hundred thousand years for a star like the Sun. This is due to the fact that each He burning reaction releases less energy than each H burning one. One needs more reactions, hence more He destruction, to obtain the same amount of energy. This effect results in the fact that the region of red giants in the HR diagram is less crowded than the Main Sequence (Figure II.5). When He is completely destroyed in the center, and if the star is massive enough, it goes on burning in the outer layers while the center begins to contract again.

---

* The H nuclear burning occurring in a star of m solar masses lasts ~ $10^{10}/m^3$ yr (with a lower limit of 10^6 yr).

The same scenario takes place once more: C burning at $\sim 8 \times 10^8$ K, and O burning at $\sim 10^9$ K. Then the star appears in the giant or supergiant region of the HR diagram. If the gravitational contraction is able to raise the central temperature to ~ 4 to 5×10^9 K, Si, which is then the most stable element, undergoes a quasi-equilibrium fusion which transforms it into Fe group elements.

The most important parameter in the study of stellar evolution is the mass. The lifetime of the star depends on it. According to the mass luminosity relation, the luminosity (i.e. the energy radiated per second) increases with m^3. Massive stars are spendthrift compared to light ones! On the Main Sequence, the internal temperature adjusts itself so that the nuclear energy production equals exactly the energy losses; the H core is exhausted much quicker in more massive stars. In the same way, the phases of gravitational contraction are more rapid in more massive stars. Another effect is that in the case of the less massive stars the gravitational energy may be too small for the whole set of processes described above to take place (Table II.2). For example, stars less massive than 0.7 solar masses can never have a central temperature high enough for C to burn. Furthermore, the final stages of stellar evolution (the death of the stars) depend on their masses. After the last nuclear phase, low mass stars slowly lose part of their external layers (they go through the 'planetary nebulae' phase while their core goes on contracting and slowly cooling: then they become white dwarfs. In cases of higher masses the core collapses while the external layers explode – it gives a supernova, one billion more luminous than an ordinary star. The exact mass limit between white dwarfs and supernovae progenitors depends on the mass loss phenomena. It is now supposed to range between 3 to $8 M_\odot$. All throughout the book, it will be seen that late stages of stellar evolution play a major role in nucleosynthesis.

TABLE II.2

The last nucleosynthetic event in stars according to their masses

Range of stellar masses (in solar mass units)	Last nucleosynthetic event
<0.1	none (black dwarfs)
0.1–0.4	H burning
0.4–0.7	He burning
0.7–0.9	C burning
>0.9	O burning

References

Quoted in the text:

Gamow, G.: 1949, *Rev. Mod. Phys.* **21**, 367.
Harrison, E. R.: 1968, *Physics Today*, June 68, p. 31.
Hertzsprung, E.: 1911, *Astron. Nachr.* **189**, 255; **190**, 119 (in German).

Hubble, E.: 1929, *Wash. Nat. Ac. Proc.* **15**, 168.

Lemaître, G.: 1931, *Revue des questions scientifiques*, quoted in Harrison (1968).

Lemaître, G.: 1949, *L'hypothèse de l'atome primitif. Essai de Cosmogonie*, Neuchâtel, Editions du Griffon (in French).

McCrea, W. H. and Milne, E. A.: 1934, *Quart. J. Math., Oxford Ser.* **5**, 73.

Omnès, R. and Puget, J. L.: 1974, *IAU Symp.* **63**, 335.

Penzias, A. A. and Wilson, R. W.: 1965, *Astrophys. J.* **142**, 419; **72**, 315.

Russell, H. N.: 1913, *Proc. Am. Philos. Soc.* **51**, 569

Steigman, G.: 1976, *Nature* **261**, 479.

Further readings:

Clayton, D. D.: 1968, *Principles of Stellar Evolution and Nucleosynthesis*, McGraw-Hill Book Co., New York.

Cox, J. P. and Giuli, R. T.: 1968, *Principles of Stellar Structure*, Gordon and Breach Science Pubs., Inc., New York.

Iben, I.: 1974, *Ann. Rev. Astron. Astrophys.* **12**, 215.

Omnès, R.: 1970, *Introduction à l'étude des Particules Elémentaires*, Ediscience, Paris (in French).

Reeves, H.: 1964, *Stellar Evolution and Nucleosynthesis*, Gordon and Breach Science Pubs., Inc., New York.

Schwarzschild, M.: 1965, *Structure and Evolution of Stars*, Dover Pubs., Inc., New York.

Sciama, D. W.: 1975, *Modern Cosmology*, Cambridge University Press, Cambridge.

Shklovskii, I. S. and Sagan, C.: 1966, *Intelligent Life in the Universe*, Dell Publ. Co., Inc., New York.

Weinberg, S.: 1977, *The First Three Minutes*, Basic Books, Inc., New York.

CHAPTER III

THE CHEMICAL COMPOSITION OF
THE OBSERVABLE UNIVERSE

One of the basic facts that nuclear astrophysics tries or hopes to explain is the chemical composition of the observable universe, namely the relative proportion of the chemical elements classified a century ago by the Russian chemist Mendeleev. Before displaying the main trends of this chemical composition, let us briefly recall the main techniques which are used by the astrophysicists to determine this composition.

III.1. Techniques for Abundance Determination

The techniques which are used for abundance determination can be divided into (1–1) the direct methods in which a given sample of material is collected and analyzed, and (1–2) the indirect methods where the composition of the matter is determined through its outcome such as the radiation it emits or absorbs.

III.1.1. THE DIRECT METHODS

To apply such methods an actual way of collecting matter must operate. For the moment this applies to four different pieces of material: terrestrial samples, lunar samples which have been collected either by astronauts or by automatic methods, meteoritic samples, and cosmic rays. For all except cosmic rays, all the material comes from a tiny part of the Solar System, namely the Earth, now the Moon and soon Mars and the meteorites. The techniques which are used for analyzing these samples are sometimes difficult to apply but are now rather well known and straightforward. Among the different techniques let us quote for instance radiochemistry and the mass spectrometry which allows us to analyze the composition of a sample by separating the nuclei with respect to their charge and/or their mass.

For cosmic rays which are very energetic particles (MeV) either of solar origin (solar Cosmic Rays) or coming from other galactic sources such as supernovae (Galactic Cosmic Rays), the way to analyze them is to use nuclear physics techniques such as nuclear spectrometers, solid state detectors, photomultiplier techniques or Cerenkov detectors (Figure III.1).

III.1.2. THE INDIRECT METHODS

Under this label are listed the methods based on the analysis of radiation emitted

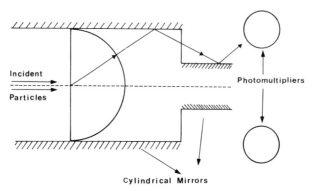

Fig. III.1.   The principle of Cerenkov detectors. In a refractive medium the velocity of light is reduced by a factor $1/n$ where n is the refractive index of the medium. It is thus possible that the velocity of a relativistic particle, close to the velocity of light in vacuum, exceeds the velocity of light in the medium. In this case the particle emits radiation in its direction of motion. This effect is used as a detection process for high energy particles. The figure (adapted from Marshall, 1952) shows the principle of one of the simplest Cerenkov detectors. The high energy particles penetrate a plexiglass lens. The emitted light is reflected by two mirrors and then magnified by a photomultiplier which is, by its position, protected from the incident particle flux.

and absorbed by the different atoms constituting a given sample of matter. The most important is spectroscopy.*

Spectroscopy is based on atomic and molecular physics: since N. Bohr, atomic physics is rather correctly understood in terms of quantum mechanics: electrons rotate around the nucleus on electronic orbits different from one another according to the species considered; electrons can jump up or down from one orbit to the other according to the absorption or emission of a photon having a *very specific* wavelength given by:

$$E_2 - E_1 = h\nu = hc/\lambda,$$

where E_2 and E_1 are the final and initial energy of the electrons and ν and λ the frequency and the wavelength of the radiation (absorbed if $E_2 > E_1$ and emitted if $E_2 < E_1$) h and c are respectively the Planck constant and the light velocity. As a consequence, a chemical species absorbs or emits light at definite wavelengths which characterize this species.

Therefore, since the light absorbed or emitted by a given star or galaxy can be analyzed as a function of the wavelength, it is possible to extract some information on the chemical composition of this star or this galaxy. Classically the spectroscopist analyzes the light coming from a given object (or absorbed by it) by means of gratings. He then obtains plates where lines or features of different intensities can

---

* New techniques of photometric observations based on the measurements of magnitude differences in distinct spectral ranges visual, Infra Red, Ultra Violet, . . . allow in some conditions to determine indirectly element abundances.

be seen. The relative abundances of the atoms in the material can be derived from the relative intensities of the lines. Contrary to what we would suppose at first hand, the method is not as straightforward as one would like. The spectroscopic methods can be summarized in the following way: the plate presenting the spectrogram of a given object is analyzed by microdensitometry techniques which give for every observed wavelength the amount of radiation collected by the plate. This spectrogram is then calibrated: one defines the position of the continuum radiation (radiation which in first approximation corresponds to a black body radiation defined thermo-dynamically by the temperature of the emitting or absorbing gas). One can thus obtain the *line profiles* (Figure III.2).

These spectroscopic techniques have largely improved over the last few years. Instead of photographic plates, astronomers now use electronic devices which allow them to observe fainter and fainter astronomical objects. The light is received on a grid of diodes which liberate electrons when they are let by photons. The electronic fluxes are then amplified through a series of photomultipliers and received on a TV screen. The spectrum is instantaneously analyzed and displayed on an oscillo-scope. It can also be printed for future analysis. Michelson and Perot-Fabry inter-ferometers are also widely used, e.g. in IR spectroscopy: In the so-called heterodyne techniques the signal is mixed with the one emitted by a local oscillator and the resulting signal is further amplified.

If, as said before, the absorption or the emission occurred exactly at a specific wavelength, the line profiles would reduce to single lines. In fact, the line profiles are broadened for three basic reasons.

(1) The *natural broadening* of the line corresponds to the fact that the transitions between electronic orbits do not correspond to a single energy but to an interval $E_0 + \Delta E_0$ (where ΔE_0 is deduced from Heisenberg uncertainty principle from the lifetime of the levels).

(2) *The Doppler broadening* comes from the thermal motion of the atomic species in the gas. Part of the atoms move towards the observer, and the corresponding radiation wavelength is then shorter while another part of the atoms (statistically) moves off (longer radiation wavelength).

(3) Finally, when there are many atoms per unit volume, they collide with one another, the effect of which being also to modify the radiation wavelength (*colli-*

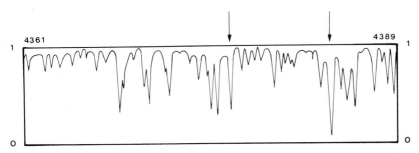

Fig. III.2.   The tracing of a stellar spectrum. This is the spectrum of Procyon (α Canis Minoris) between 4361 and 4389 Å. The two arrows show the Fe I lines at λ4375.93 and 4383.57.

sional broadening). From these line profiles, one can derive the abundance of a given chemical species. The profile of the line is traditionally characterized by its *equivalent width*, which is defined as the width of the rectangle which has the same area as the line profile and a length equal to the intensity of the continuum (Figure III.3).

Suppose fcʳ a moment that there is a magician able to change the abundance N of an element in a stellar atmosphere. He begins with a very small change and increases it little by little. He also measures the equivalent width of a given line of this element and watches it varying with N. It first varies proportional to N. But then it stops varying even with increasing N and after a while, it varies again, but now like $\sqrt{N}$. You obtain in that way the so-called theoretical curve of growth (Figure III.4a). The reason for this behavior is simple: first the atoms absorb more and more light when their number increases and the depth of the line increases as well. But when all the available light is absorbed in the line, its depth cannot increase anymore the line is *saturated*. Its width, which is then due to Doppler broadening, does not change with increasing abundance until the atoms are so numerous that the collisional broadening becomes important. The theory of collisional broadening shows that in this case the width increases like $\sqrt{N}$.

From Figure III.4a, we could infer that a simple measure of the equivalent width of a line in a star would give the abundance of the element. This is not as simple because the curve of growth is not unique; it is actually different for each element and for each star, as it depends on the physical conditions in the stellar atmosphere. We are faced with this apparently insoluble problem: to construct the curve of growth for an element in a star before being able to measure its abundance. It is like a cat trying to bite is tail. Fortunately nature provides the solution. Each element produces several different lines, with different strengths. The strength of a line, characterized by the so-called 'oscillator strength f', is proportional to the probability for the photon to be emitted or absorbed at the wavelength of the line. The variation

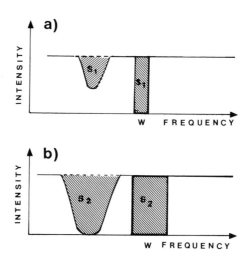

Fig. III.3.   The equivalent width is the width W of a rectangle which has the same area as the line surface, and a length equal to the intensity of the continuum. (a) Non saturated line. (b) Saturated line.

of f from line to line can be used instead of the variation in N to construct the curve of growth. Then the ratio W/λ, where λ is the wavelength of the line, is plotted instead of W in ordinates. One obtains in this way the so-called empirical curve of growth (Figure III.4b) which allows the astronomers to derive stellar abundances with an uncertainty of a factor 2.

Spectra of interstellar clouds and galaxies also give information on their composition. In the case of molecules, the lines correspond to changes in their rotation or vibration. They generally emit or absorb at radio wavelengths. From the profiles of these lines, one can deduce for example the number of molecules in the line of sight in the case of a thin cloud, the temperature of the cloud, etc. .... There are an increasing number of observations in radioastronomy which provide more and more information on our Galaxy and other galaxies.

Besides the optical and radio domains, the modern techniques allow us to investigate the other wavelength ranges. The most energetic photons correspond to the range of γ rays ($E >$ a few MeV), X-rays ($E >$ a few keV) and the UV. The photons which are less energetic than optical photons emit in the IR and the millimeter range up to the radio wavelengths. These different wavelength domains correspond to different observation techniques. The γ ray, X-ray astronomy and part of the UV and IR astronomy can only be done on space platforms (the Earth's atmosphere either absorbs or diffuses these photons (see Table III.1). More and more sophisticated equipment for observing radiation at such wavelengths is launched in balloons, in satellites and around 1980 will be put in 'out of atmosphere laboratories' (Spacelab, Space Telescope, Shuttle . . .).

These various electromagnetic manifestations can be used to obtain information on the chemical abundance of the Universe. Let us quote for example the observations of γ ray lines characteristic of the excited states of the nuclei of given elements, X-ray absorption lines due to metals, UV lines produced by the absorption of light coming from hot stars, IR and mm lines characteristic of molecules . . . Spatial, radioastronomical, IR and optical techniques are presently making so much progress that we believe we are living in one of the golden ages of astronomy.

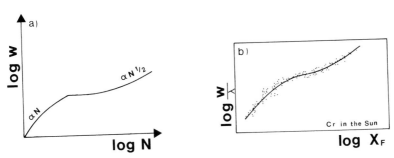

Fig. III.4.   Curves of growth: (a) Theoretical curve of growth. For weak lines the equivalent width varies like N, for Doppler lines it is constant, and for saturated lines it varies like $\sqrt{N}$. (b) Empirical curve of growth, for Cr in the Sun. Log W/λ is plotted versus a function of the oscillator strength.

TABLE III.1

λ	ν or $h\nu$	Type of radiation	Observations from				Detectors
			Earth	Plane	Balloon	Satellites	
10^{-5} Å	1240 MeV						Geiger counters
10^{-4} Å		γ Rays					Scintillators
10^{-3} Å	12.4 MeV						
10^{-2} Å							Cerenkov light
10^{-1} Å		X-Rays					detectors
1 Å	12.4 eV						
10 Å							
100 Å	3×10^{16} s^{-1}	UV					Photographic
1000 Å							Photoelectric
10 000 Å = 1 μm	1.24 eV	Visible	Visible window				Eye
10 μm							Photoconductiv
100 μm	0.012 eV	IR					detectors
1000 μm = 1 mm							
10 mm = 1 cm	30 000 MHz						Bolometers
10 cm		UHF					
100 cm = 1 m	300 MHz		Radio window				
10 m		Short wave					Radio-receivers.
100 m	3 MHz						
1000 m = 1 km							
10 km		Long wave					
100 km	3 kHz						
1000 km							

III.2. The Abundances of the Elements in the Universe

From the direct determination of the abundances in the Solar System (one generally uses a specific sample, the carbonaceous chondrites which are believed to be the most primitive material of the Solar System) and the indirect determination of the solar abundances, one defines a set of abundances of this curve now called the Standard Abundance Distribution (SAD) (Figure III.5). The main characteristics of this curve are the following:

(1) H and He are the most abundant elements in the observed Universe. They represent $\sim 97\%$ of the total mass. $n(\text{H})/n(\text{He}) \sim 10$ almost everywhere.

(2) When the atomic mass increases the abundance decreases very sharply up to $A \simeq 50$. One must note however that the light elements Li, Be and B are much underabundant with respect to their neighbors (Li Be B/CNO $\sim 10^{-5}$); and the nuclei with an even atomic mass are generally more abundant than the odd nuclei. The elements whose nucleus can be considered as constituted of α particles (two protons + two neutrons) are also relatively more abundant than their neighbors. After H and He the most abundant nuclei are O and C.

(3) The abundance curve shows a peak at masses $50 < A < 70$ with a maximum for ^{56}Fe.

(4) The abundance curve then decreases again for the highest atomic masses

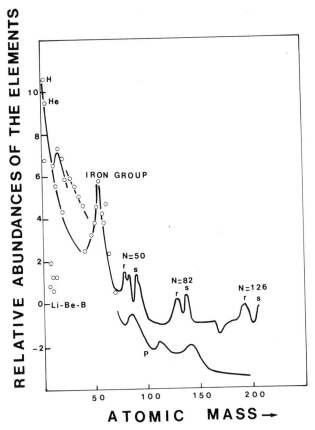

Fig. III.5.   The standard abundance distribution of the elements in the Universe.

but much more slowly. The curve is not monotonous. There exists less pronounced peaks called magic peaks (Chapter VI). Furthermore, the abundance curve can be split into two parts; the upper part gathers the elements with neutron rich nuclei; while the lower part gathers those with proton rich nuclei. The purpose of the nucleosynthesis and therefore our purpose in the next chapters is to find a consistent explanation for these striking characteristics.

It is important to realize that a large number of stars and galaxies, as well as the interstellar matter, follow the SAD distribution curve. However, there are important and interesting departures from it. A majority of them may be explained by nuclear effects and are often related to the chemical evolution of galaxies (Chapter IX). Some others are due to physical effects:

(1) A correlation can be made between the abundances of the metals,* ($Z \geq 6$), and the location (and the age) of the stars. The older stars which belong to the halo

* Do not confuse the chemical definition and the astronomical definition of metals (for astronomers, C, N, and O are metals!

of our Galaxy and are members of the globular clusters, have a metal abundance lower by a factor 10 to 10^3 than the solar metal abundance. Theories on chemical evolution of galaxies are presently trying to account for this relation between the age of a star and its metallicity.

(2) In galaxies (such as ours) the metallic abundance varies with the distance to the galactic center. The metal abundance increases when one goes from the outer regions into the center of these galaxies. Theories on chemical evolution of galaxies also take into account these abundance gradients in nearby spiral galaxies.

(3) Among different galaxies, the metal content seems to vary; for instance, the two satellites of our Galaxy, the Small and the Large Magellanic Clouds (SMC and LMC) have a metallic abundance lower than that of our Galaxy. The effect is more pronounced for the SMC than for the LMC. This seems to be related to a lower rate of star formation (Chapter IX) for these two galaxies.

(4) The isotopic composition of the matter is only known for the Solar System (especially in the Earth and meteorites).* For instance, the abundances of the heavy elements with proton rich nuclei can only be determined in the solar system. Now, the situation has improved in the sense that for the lightest elements isotopic ratios have been observed in cool stars and in the interstellar medium. As an example, the D abundance has been measured in various spots of the interstellar medium and the D/H ratio has been found to be 10 times lower than in the Earth and the carbonaceous chondrites. The $^{13}C/^{12}C$ ratio has been measured in the interstellar medium and in the atmosphere of cool stars and has been found 2 to 4 times higher in the interstellar medium and 4 to 20 times higher in some cool stars than in the Solar System. We will come back to these interesting features in Chapter IX.

(5) Up to recently, it seems that the Solar System itself presented a very homogeneous isotopic composition suggesting a simple origin. There is growing evidence that this picture is far too simple. New discoveries of isotopic anomalies in O (Clayton et al., 1973), in Ne (Black, 1971, Eberhardt, 1974), and in Mg (Lee et al., 1976), led to the conclusion that the whole Solar System material has not a single origin. This point will be analyzed in more detail in Chapter VIII.

(6) When one studies the galactic cosmic ray composition and compares it to the solar abundances one notes that metals are enriched with respect to H and He by factors 10 to 100; moreover, the light elements Li, Be, B are 10^4 more abundant than what is observed in the stellar surfaces (Figure III.6). This may be explained in terms of spallation reactions (Chapter VII).

(7) The abundances in the nearby interstellar medium have been thoroughly studied by UV astronomy (the Copernicus spatial experiment launched by the Princeton Observatory group). From present available data it would seem that the metallic abundances are weaker than the solar values. This has been interpreted in terms of the presence of dust in the interstellar medium in which the most refractory metals are preferentially trapped (Chapter II).

---

* Some isotopic abundances have actually been measured in the interstellar medium (H, C, N, O, and S), in some red giant stars (C and O) and in peculiar stars (He in 3 Cen A and related stars, Hg and Pt in Hg-Mn stars. See next section for a definition of these stars).

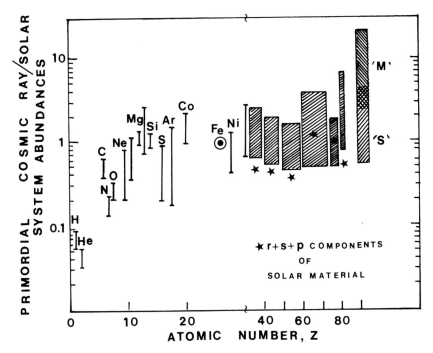

Fig. III.6.   Element abundances in cosmic rays in units of the solar abundances.

(8) Important groups of stars show metallic abundances very different from the SAD ones. Let us quote the 'supermetallic stars' with a very high metal content, the 'Ba stars', or giant stars with abnormally strong Ba abundances, the C, N, and O stars, with very intense C, N, and O features, the He rich stars, the He weak stars, the 'δDel' stars, with low Ca II lines and variable luminosity, and the well-known group of the so-called 'peculiar stars' and 'metallic stars' which may be explained by physical processes in the envelope (next section).

(9) Finally, we must mention the unique but fascinating object FG Sagittae.* Since 1894, the brightness of this star which is surrounded by a nebulosity of plane-tary nebulae type, has steadily increased in luminosity by a factor of about 40, up to 1968, then slightly decreased. Meanwhile, the star has been observed to cool off at a rate of about 250 K yr^{-1} going from A to F spectral type. At the same time anomalous strong absorption lines of Y, Zr, Ce, and La appeared in the spectrum. They could be a signature of nucleosynthetic events (the s process defined in Chapter VI) which happened recently. FG Sge could actually be in the process of ejecting a new planetary nebula shell, and the presently observed nebulosity could be an older by-product of the same process.

---

* Another star CI Cygni may be similar in this respect to FG Sagittae.

III.3. Main Sequence Peculiar Stars*

Main Sequence peculiar stars are found in a rather well defined range in the HR diagram, which corresponds to spectral types B-A-F. They are usually classified into two groups: (1) Ap stars, or 'peculiar stars', and (2) Am stars, or 'metallic line stars'. These two groups of stars differ from one another by their general physical properties (for example, Ap stars are hotter than Am stars), and by the order of magnitude of their abundance anomalies (one order of magnitude for Am stars, three or more for Ap stars).

Ap stars do not represent a well defined class. It has to be separated into two sub-classes: the magnetic Ap stars, in which magnetic fields of more than 10^3 G have been observed, (e.g. the 'Si stars') and the non-magnetic Ap stars (e.g. the 'Hg-Mn stars') in which no magnetic field has been observed. Within these sub-classes, Ap stars generally show large overabundances of one or several metals, which vary from one star to the other. Hence, the Ap group is accordingly divided in other different subsets corresponding to the most important anomaly. One of the most important characteristics of peculiar stars is that they all are slow rotators, which means that their atmospheres are more stable than that of normal stars. An important difference between Am stars and non-magnetic Ap stars is that the percentage of binary systems among Am stars is nearly 100% while it is less than 30% among non-magnetic Ap stars. The possible variability of the stellar luminosity is also important: the Am and non-magnetic Ap stars are not variable, while the magnetic Ap stars are. On one hand, Am stars lie in the same region of the HR diagram as δ Scuti stars which are of similar spectral type but with a variable luminosity. It seems that the two stellar classes Am, δ Scuti exclude one another: these stars either have a constant luminosity but an abnormal chemical composition (Am phenomenon), or their luminosity is variable but their chemical composition is normal (δ Scuti phenomenon). On the other hand, for the same region of the diagram, magnetic Ap stars have a variable luminosity while normal A stars have a constant one. Moreover, magnetic Ap stars are not only variable in luminosity but the *profiles* (or shapes) of the observed lines also vary with time. Figure III.7 shows the position in the HR diagram of Ap and Am stars. As pointed out by Jaschek and Jaschek (1958), and Sargent and Searle (1967), abundance anomalies are related to the surface temperature of the stars inside the two sub groups of magnetic and non-magnetic Ap stars.

In summary, information on Main Sequence peculiar stars has increased significantly these last few years so that their characteristics are now rather well established. This implies the choice for possible mechanisms which can be invoked to explain such anomalies.

The presently accepted theory is that the abundance anomalies observed in Ap and Am stars are the result of diffusion processes in their outer layers (Michaud, 1970). When the stellar outer layers are stable against turbulence and convection, elements are subject to several forces which make them diffuse with respect to H (the main constituent of the stellar gas). The two basic forces are the force due to

---

* For a complete review of Main Sequence peculiar stars, see Preston (1974).

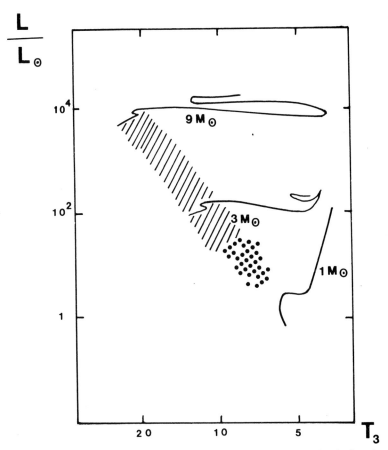

Fig. III.7. Location of the HR diagram of the Main Sequence peculiar stars. The shadowed area is for Ap stars while the dotted area is for Am stars, T_3 represents the temperatures of the stars in thousands of degrees.

the gravity (or the hydrostatic pressure gradient) which makes heavy particles diffuse towards the stellar center, and the force due to the radiation pressure, which makes particles diffuse towards the stellar surface. Whether particles diffuse downwards or upwards depends on the relative intensity of these two forces. As a general rule, the radiation force on metals is higher than the gravitational force in Ap and Am stars, so that metals diffuse outwards. This is not true when the metal is ionized in a complete shell configuration in which case the radiation force is weak so that the metal diffuses inwards (this is, for instance, the case of Ca in Am stars).

In the frame of the diffusion theory, abundance anomalies in Ap stars are well accounted for if the atmosphere is stable against convection so that diffusion may happen there, and abundance anomalies in Am stars are well accounted for if diffusion occurs under the outer convection zone.

As a conclusion, we can say that nearly all the observed departures from the SAD curve have now found self consistent explanations. We are still left with the task of explaining the characteristics of the SAD curve itself. As stated before, all these exciting and recent discoveries make nuclear astrophysics a lively field. The purpose of the subsequent chapters is to try to give a flavor of it.

References

Quoted in the text:

Black, D. C.: 1971, *Geochim. Cosmochim. Acta* **35**, 230.
Clayton, R. N., Grossman, L., and Mayeda, T. K.: 1973, *Science* **182**, 485.
Cowley, C. and Cowley, A.: 1964, *Astrophys. J.* **140**, 713.
Eberhardt, P.: 1974, *Earth Planet. Sci. Letters* **24**, 182.
Hiltner, W. A. and Williams, R. C.: 1946, *Photomatic Atlas of Stellar Spectra*, University of Michigan Press, Ann Arbor.
Iben, I.: 1967, *Ann. Rev. Astron. Astrophys.* **5**, 571.
Jaschek, C. and Jaschek, M.: 1958, *Z. Astrophys* **45**, 35.
Lee, T., Papanastassiou, D., and Wasserburg, G. J.: 1976, *Geophys. Res. Letters* **3**, 109.
Marshall, J.: 1952, *Phys. Rev.* **86**, 685.
Michaud, G., 1970, *Astrophys. J.* **160**, 641.
Preston, G. W.: 1974, *Ann. Rev. Astron. Astrophys.* **12**, 257.
Sargent, W. L. W. and Searle, L.: in R. C. Cameron (ed.), *The Magnetic and Related Stars*, Proceedings of a NASA Symposium, Monobook Corp., Baltimore.
Trimble, V.: 1975, *Rev. Mod. Phys.* **47**, 877.

Further readings:

Audouze, J. (ed.): 1977, *CNO Isotopes in Astrophysics*, D. Reidel Publ. Co., Dordrecht, Holland.
Burbidge, E. M., Burbidge, G. R., Fowler, W. A., and Hoyle, F.: 1957, *Rev. Mod. Phys.* **29**, 547.
Cameron, A. G. W. (ed.): 1973, *Cosmochemistry*, D. Reidel Publ. Co., Dordrecht, Holland.
Jefferies, J. T.: 1968, *Spectral Line Formation*, Blaisdell, Waltham.
Mason, B. (ed.): 1971, *Elemental Abundances in Meteorites*, Gordon and Breach Science Pubs., Inc., New York.
Michaud, G.: 1975, in W. W. Weiss, H. Jenker, and H. J. Wood (eds.),'Physics of Ap Stars', *IAU Colloq.* **32**, Vienna Observatory, Vienna.
Mihalas, D.: 1970, *Stellar Atmospheres*, W. H. Freeman and Co., San Francisco.
Peimbert, M.: 1975, *Ann. Rev. Astron. Astrophys.* **13**, 113.

CHAPTER IV

THERMONUCLEAR REACTIONS AND NUCLEAR REACTIONS IN STELLAR INTERIORS

IV.1. Nuclear Reactions: Generalities

When a nucleus a hits a target A, their interaction can lead to different products, each possibility being referred to as a 'channel'. All throughout this book we assume that the target nucleus is the heavier nucleus. The effects we are discussing here are symmetrical and remain the same if the target is the lighter one. When the final products of the nuclear interaction are identical to the initial ones, it is called 'scattering'. The scattering is said to be elastic when the respective energies (or velocities) of the interacting particles are not modified by the interaction, and inelastic when the respective energies are modified by it. In the other cases, the interaction is a nuclear reaction:

$$A + a \to A + a \text{ (scattering)}$$

$$\left. \begin{array}{l} A + a \to B + \dots \\ A + a \to C + \dots \\ A + a \to D + \dots \end{array} \right\} \text{nuclear reactions.}$$

When the mass of the heavy product B is larger than the mass of the target A, the nuclear reaction becomes a fusion reaction. Otherwise, it is a fission reaction (or spallation reaction: see Chapter VII).

Nuclear reactions either release or absorb energy according to the difference in the binding energies of the final vs initial nuclei. The mass of a given nucleus is less than the sum of the mass of its components. Take for example the Fe nucleus, ^{56}Fe, which is made of 26 protons and 30 neutrons. The binding energy per nucleon of this nucleus is defined as:

$$\frac{1}{56}(26M_p + 30M_n - M_{56})c^2,*$$

where M_p is the mass of one proton, M_n the mass of one neutron, M_{56} the mass of one Fe nucleus, and c the velocity of light. The masses are generally given in MeV, in which case one has to drop the c^2 factor to get the binding energy also in MeV. Figure IV.1 gives the binding energies per nucleon of the nuclei as a function

---

* In the general case, the binding energy of the nucleus with an atomic mass A, atomic number Z and mass M_A is given by:

$$\frac{1}{A}(ZM_p + (A - Z)M_N - M_A)c^2.$$

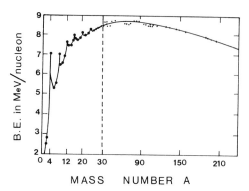

Fig. IV.1. Binding energy of the nuclei in MeV per nucleon, as a function of their mass number A. Note the sharp peaks corresponding to the magic numbers of nucleons (see text), and the maximum for ^{56}Fe. The scale changes at $A = 30$.

of their mass. ^{56}Fe is the most bound nucleus. The peaks correspond to the ^{4}He nucleus (also called the α particle) which is tightly bound, and to its multiples ^{12}C, ^{16}O, and ^{20}Ne (Be is unstable). This figure shows that the nucleosynthesis of the elements lighter than ^{56}Fe is quite different from that of the heavier elements: the fusion reactions implying the light elements are releasing energy (they are called exoenergetic) while those implying the heavier ones absorb it (endoenergetic). The energy released or absorbed by the fusion reaction $A + a \rightarrow B + b$ is equal to $Q = B_B + B_b - B_A - B_a$. If Q is positive, the reaction transforms less bound nuclei to more bound nuclei and is exothermic. If Q is negative, the reaction is endothermic and in this case Q is called the threshold of the reaction. The binding energies of nuclei are generally given in terms of 'atomic mass' excess; it provides a suitable way to compute the Q values of the reaction (see Appendix B).

The present chapter is devoted to the fusion reactions synthetizing the light elements which constitute the main source of the stellar energy. From Figure IV.1, we can see that the binding energies increase largely in the range of the lightest nuclei and level off for $A \geq 26$. The energy released is much more important in the fusion reactions implying the lightest nuclei ($A < 12$) than in those implying the heaviest ones. As a consequence, the time scale of the first fusion phases is much larger than the time scales of the subsequent fusion periods.

The total energy production by nuclear reactions, as well as the abundance changes, is directly related to the number of reactions which can occur per unit time, i.e. the nuclear reaction rates.

IV.2. Nuclear Reaction Rates

The probability for a given reaction between two nuclei A and B to occur is characterized by a physical quantity which has the dimension of a surface: *the reaction cross section*. This cross section is similar to the surface of the target used for a dart-game: the larger the target, the higher the probability for the player to hit it. In the same way the probability for the nuclear reactions between the incident nucleus ('the dart') and the target nucleus is higher for a larger cross section. The cross section is exactly defined as the ratio of the number of reactions (per target nucleus and per time unit) to the flux of incident particles (i.e., the number of incident particles per surface unit and per time unit):

$$\text{cross section} = \frac{nb \text{ of reactions}/nb \text{ of part. } A/\text{time unit}}{nb \text{ of part. } a/\text{surface unit}/\text{time unit}},$$

where A is the target and a the projectile. The cross section $\sigma_{A,a}(v)$ depends on the relative velocity of the interacting nuclei A and a, which is defined as the difference between their two velocities. The characteristic cross section units are the barn (10^{-24} cm^2) and the millibarn (10^{-27} cm^2).

We can now define the nuclear reaction rates per unit volume. Let us consider the simple case of a gas of nuclei A (with N_A nuclei per volume unit) bombarded by a flux of high velocity particles a. We suppose that the particles A are 'at rest' $(v(A) = 0)$. In this case, the nuclear reaction rate is the product of $\sigma_{A,a}(v)N_A$ and the flux of nuclei a, which is simply equal to vN_a. Thus the reaction rate $r_{A,a}$ can be written as:

$$r_{A,a} = \sigma_{A,a}(v)N_A N_a v.$$

Inside the stars, the gas of nuclei is generally in thermodynamical equilibrium, so that the velocities of all the nuclei (and also the relative velocity between nuclei) follow the Maxwell-Boltzmann distribution law* (Figure IV.2). The expression for the reaction rate then becomes:

$$r_{A,a} = N_A N_a \langle \sigma v \rangle_{A,a},$$

where $\langle \sigma v \rangle_{A,A}$ is the product σv averaged over all the possible velocities, taking into account the Maxwell-Boltzmann distribution law. The quantity $\lambda_{A,a} = \langle \sigma v \rangle_{A,a}$ is referred to as the nuclear reaction rate per pair of particles. The *lifetime* of a nucleus A against the reaction (A, a) is defined as**:

$$\tau_{A,a} = \frac{1}{\lambda_{A,a} N_a}.$$

The number of nuclei A which are destroyed within a time Δt is given by:

$$\Delta N_A = r_{A,a}\Delta t = \lambda_{A,a}N_A N_a \Delta t = \frac{N_A}{\tau_{A,a}}\Delta t.$$

Let us shortly indicate how $\lambda_{A,a}$ can be evaluated. Figure IV.3 shows the general

---

* The number of nuclei with a velocity between v and $v + dv$ is:

$$N(v)\,dv = N\left(\frac{m}{2\pi kT}\right)^{3/2} \exp\left(-\frac{mv^2}{2kT}\right)dv.$$

** If the nucleus A is involved in several nuclear reactions, its total lifetime τ_A is given by:

$$\frac{1}{\tau_A} = \sum \frac{1}{\tau_{A,a}}.$$

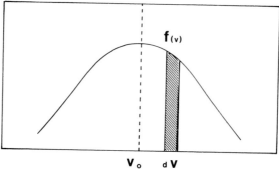

Fig. IV.2. The Maxwell-Boltzmann distribution law. The shaded area represents the number of particles which have a velocity between v and $v + dv$.

behavior of a nuclear reaction cross section as a function of the energy of the incident particle. Different energy ranges can be distinguished: at low energies ($E <$ some 100 keV) the cross section increases rapidly and continuously with E. At larger energies ($E <$ some tens of MeV) the curve shows peaks which are called resonances while the cross section does not increase anymore with E (the physical explanation of the resonances will be given later on in this chapter). At still larger energies, the resonances are so numerous that they overlap, leading to a smooth curve, but also to a much larger cross section.

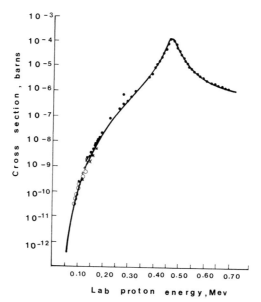

Fig. IV.3. Experimental (dots and crosses) and theoretical (solid line) cross section for the nuclear re-action $^{12}C + p \rightarrow {}^{13}N + \gamma$ vs the energy of the incident proton. The resonance occurs for a proton of 0.406 MeV.

The probability that a reaction occurs between two charged nuclei depends on two different kinds of physical interactions: (1) they have to overtake the electrical repulsion due to their charge (or Coulomb repulsion); and (2) once they are close enough (inside a few 10^{-13} cm) they interact, with a probability depending on purely nuclear parameters. The repulsive potential between two nuclei of charge Z_A and Z_a is given by the Coulomb law $(Z_A Z_a e^2)/r$. The Coulomb barrier $B_{A,a}$ between these two nuclei is defined as the maximum of this repulsive potential:

$$B_{A,a} \,(\text{MeV}) = 1.4 \,\frac{Z_A A_a}{R_{(\text{fm})}},$$

where R is the interaction distance of the two nuclei:

$$R = 1.3(A_A^{1/3} + A_a^{1/3}) \,\text{fm.}^*$$

This Coulomb barrier is at least a few MeV for all the interacting nuclei while the average thermal energy inside a star goes from a few keV up to a few hundreds of keV.

On the average, a nucleus inside a star does not have enough energy to overtake the Coulomb repulsion. However, once in a while, a less energetic nucleus can cross the Coulomb repulsion due to a quantum effect called *tunnel effect* (Figure IV.4). This tunnel effect is related to *the Heisenberg uncertainty principle* which expresses the fact that the position of a particle which has a definite energy is uncertain. The incident nucleus does not 'know' on which side of the Coulomb repulsion barrier it is, so that it has a chance to be close enough to the target to interact with it. This is the only way for a nuclear reaction to occur at low energies.

At such low energies, the cross section increases rapidly with energy because the nuclei have more chances to overcome the repulsive effect (see Figure IV.3). When the energy of the incident nucleus becomes larger than the Coulomb barrier, the cross section becomes nearly constant with E.

The nuclear parameters involved in a nuclear reaction are well accounted for by *the compound nucleus model*. In this model the target 'swallows' the incident nucleus

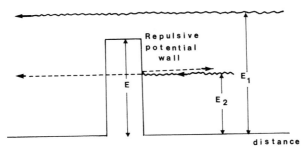

Fig. IV.4. *Tunnel effect*: From classical mechanics, a particle which finds a repulsive potential on its way would be reflected unless it has an energy E_1 larger than the energy E of the wall. From quantum mechanics we know that a particle with an energy E_2 smaller than E can indeed cross the potential wall: part of the wave associated with the particle is reflected, part is transmitted through the wall.

* 1 fm (Fermi) = 10^{-13} cm.

to form the 'compound nucleus' which further disintegrates into two or more components (the final products of the reaction). The compound nucleus can be described in the frame of the shell model. In this model the physics of the nucleus is similarly described as in the case of an atom. The nucleons in a nucleus are on quantified 'orbits' (or 'levels') as in the case of the electrons in an atom. A nucleus can be 'excited' as well as an atom (Chapter III), when one or several nucleons are on more energetic orbits than usual. The deexcitation of the nucleus (when the nucleons go back to their own 'orbits') leads to the emission of a photon. Now, if the sum of the energies of the nuclei A and B is equal to the energy of one of the excited levels of the compound nucleus (Figure IV.5) the probability of the reaction, and the cross section increases drastically; this effect accounts for the resonances in the cross section. The first resonance occurs at some tens of keV for heavy elements and some MeV for light ones; there is no resonance at lower energies. At high energies (at least a few MeV), the nuclear energy levels become very close, which leads to an overlapping of the resonances.

Let us go back to the nuclear reaction rate. This rate is obtained by multiplying the reaction cross section by the relative velocity of the nuclei and then averaging over all the possible velocities. Figure IV.6 shows how this can be done at low energies (outside resonances). As the reaction rate depends on the product of the cross section (which increases rapidly with energy) by the Maxwell-Boltzmann distribution (which decreases with energy within this energy domain), there is an optimal energy called *Gamow energy* (E_G) for which the reaction rate is maximum. At low energy (tunnel effect) the reaction cross section is well reproduced by the Gamow formula:

$$\sigma(E) = \frac{S(E)}{E} \exp - \left(\frac{E_G}{E} \right)^{1/2},$$

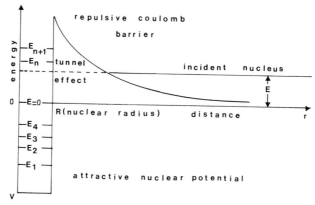

Fig. IV.5. The compound nucleus model. The compound nucleus formed by the two colliding nuclei is represented by its nuclear potential and its nuclear states $E_1, E_2, \ldots E_n, E_{n+1} \ldots$ One of the original nuclei (the target) is supposed at rest. The other one (the incident particle) has an energy E. If E does not coincide with one of the nuclear states of the compound nucleus (which is the case in the figure), the cross section is relatively small. It increases by several orders of magnitude if E coincides with one of the energies E_n (resonance effect).

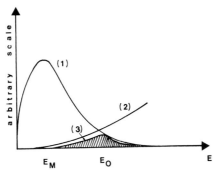

Fig. IV.6. *The Gamow peak*: The nuclear reaction rate (curve 3) is the product of the nuclear reaction cross section (curve 1) by the relative velocity of the nuclei, which, for a perfect gas, follows a Maxwell-Boltzmann distribution (curve 2). The nuclei which have the maximum probability of undergoing a nuclear reaction are *not* the ones with the average maxwellian velocity (energy E_M), but the ones with the relative energy E_0, corresponding to the maximum of the Gamow peak.

where $S(E)$ is a slowly varying function of the energy which depends on the nuclear properties of the reaction. The cross section varies also in $1/E$ as the geometrical cross section does (proportional to the square of the de Broglie wavelength) the probability for the nuclei to cross the Coulomb barrier by tunnel effect is exp $-(E_G/E)^{1/2}$. Let us note that the Gamow energy increases with temperature for a given reaction. Resonances must be taken into account when they lie close to the Gamow energy. In this case, the average of the product $\langle \sigma v \rangle$ is equal to its value at the energy of the resonance. At the resonance the cross section σ is approximated by the Breit-Wigner formula (see for example Clayton, 1968). The reaction rate is then very high within the narrow temperature range corresponding to this resonance.

IV.3. Hydrogen Burning

The most important reactions leading to the transformation of H into He inside a star are listed in Table IV.1. We will discuss separately the three basic 'chains' of H burning.

IV.3.1. THE PROTON–PROTON CHAIN, OR PPI CHAIN

Since ^{2}He does not exist in nature, the reaction between two H nuclei necessarily involves the transformation of one proton into one neutron, this neutron being further bound to the proton to give a D (nucleus of deuterium, or heavy H made of 1 proton and 1 neutron). This transformation is a weak interaction. It releases a positron (positive electron) and a neutrino:

$$H + H \rightarrow D + e^+ + \nu.$$

Weak reactions have cross sections much smaller than nuclear (or strong) ones. Their cross sections are measured in units of 10^{-44} cm^2, i.e. 10^{20} times smaller than the unit used for nuclear cross sections (the barn). Thus, the reaction $H + H \rightarrow D + e^+ + \nu$ is extremely slow and cannot be observed in the laboratory (we would

TABLE IV.1
Hydrogen burning in stars: The PP chains

---

1. proton – proton chain (PPI)

$$H + H \longrightarrow D + e^+ + \nu$$

$$D + H \longrightarrow {}^3He + \gamma$$

$${}^3He + {}^3He \longrightarrow {}^4He + 2H + \gamma$$

---

2. chains with a He catalyst

$${}^3He + {}^4He \rightarrow {}^7Be + \gamma$$

${}^7Be + e^- \longrightarrow {}^7Li + \nu$	${}^7Be + p \longrightarrow {}^8B + \gamma$
${}^7Li + H \longrightarrow {}^8Be + \gamma$	${}^8B \longrightarrow {}^8Be + e^+ + \nu$
(PPII)	(PPIII)

$${}^8Be \longrightarrow 2\,{}^4He + \gamma$$

---

have to watch for about 10 yr in a very powerful accelerator to see but one reaction!). However, its cross section is theoretically accurately determined; it is of about 10^{-47} cm^2 for 1 MeV protons. The lifetime of H in regions where the temperature is 15×10^6 K and the density 100 g cm^{-3} is 10^{10} yr (age of the oldest stars). The slowness of this reaction is one of the reasons why stars last so long only by burning their H. Should this reaction have not been so slow, our Sun would have already disappeared a long time ago (the same for Earth and its inhabitants!).

Deuterium (D) is very quickly transformed into ^{3}He (a He nucleus made of 2 protons and 1 neutron) by absorbing another proton. The abundance of D relative to H has an equilibrium value in the solar center of about D/H* = 3×10^{-18}. If we compare this value with the observed terrestrial one (D/H $\simeq 10^{-4}$), we become aware that an explanation of the observed D is not straightforward. We will come back to this problem in Chapter VII.

The ^{3}He nuclei then react in pairs to give a ^{4}He nucleus and two protons. This so-called PPI *chain* occurs basically at temperatures of 10 to 20×10^6 K and at densities of about 100 g cm^{-3}. It is the basic nuclear process which occurs inside the Sun. Figure IV.7 shows at the present time, the relative abundances of H, ^{3}He and ^{4}He, inside the Sun, as a function of its radius. In the center, H has been completely transformed into ^{4}He, due to nuclear reactions while it still has its original abundance on the surface. Inside the Sun, the ^{3}He formed by the (D + H) reactions is not yet destroyed by the (^{3}He + ^{3}He) one. Its abundance is maximum at a radius of about 0.6 solar radius. In the center, it has already been transformed into ^{4}He while on the surface it has not yet been formed.

---

* The symbol D/H stands for the ratio of the number of D nuclei per volume unit to the number of H nuclei per volume unit. This notation will be used throughout the monograph.

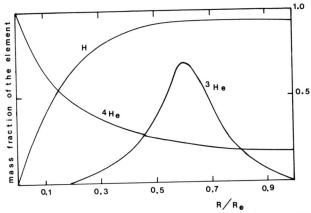

Fig. IV.7.   Abundances of H, ^{3}He, and ^{4}He vs depth in the Sun: ^{4}He is formed and H is destroyed by the nuclear reactions in the center of the Sun. The peak for ^{3}He, at $R \simeq 0.6R_\odot$ is due to the competition between the nuclear reactions which form and destroy it (see Table IV.1).

IV.3.2. THE PROTON CHAINS WITH A HE CATALYST, OR PPII AND PPIII CHAINS

If the temperature is higher than 20×10^6 K (e.g., in the center of stars more massive than the Sun) and if the ^{4}He abundance increases, He itself can act as a catalyst and induce the so-called PPII *and* PPIII *chains* (Table IV.1).

The first reaction of these chains occurs between ^{3}He and ^{4}He, giving a ^{7}Be nucleus and a photon. Then, the ^{7}Be can have two different fates. At relatively low temperature (20 to 30×10^6 K), it catches an electron and gives a ^{7}Li nucleus. The ^{7}Li reacts with H to give ^{8}Be which is highly unstable and breaks into two ^{4}He within a time of 10^{-16} s. This chain is called the PPII chain. ^{7}Be may also catch a proton and give a ^{8}B nucleus which becomes ^{8}Be by β decay and leads to two ^{4}He: this is the PPIII chain.

The disintegration of ^{8}B into ^{8}Be liberates a very energetic neutrino $E_\nu = 7$ MeV while the energy liberated by (H + H) is only ~ 0.26 MeV and the energy liberated by (^{7}Be + e$^-$) is $\simeq 0.80$ MeV. Although the dominant chain in the Sun is the PPI, the neutrinos due to PPIII should be detectable on Earth, while those due to PPII can only be marginally detected (the neutrinos due to PPI are not energetic enough). However, the huge neutrino detector (Figure IV.8) built by R. Davis, Jr., from the Brookhaven National Laboratory, with 400 000 liters of CCl$_4$ buried in a South Dakota gold mine, a few hundred meters deep, has now detected about only one fourth of the expected amount of solar neutrinos. This experiment is very important since it is the only way to measure directly the temperature in the center of the Sun: the flux of ^{8}B, and therefore, of energetic neutrinos, is a direct function of the temperature. While the photons emitted in the solar center are reabsorbed by the surrounding matter, the neutrinos, which only undergo weak interactions, can escape the Sun without being significantly absorbed. The discrepancy between the observed and the theoretically predicted neutrino flux provides one of the worst puzzles of modern astrophysics. Hundreds of ideas have been proposed to attempt to explain this discovery but none is really satisfactory.

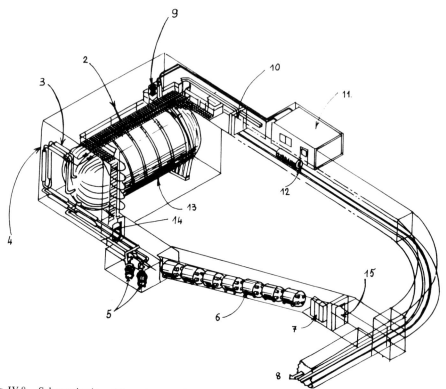

Fig. IV.8.   Schematic view of the solar neutrino detector. (1) Located inside a gold mine in South Dakota and designed by Prof. R. Davis, Jr., of Brookhaven laboratory. (2) Neutron gun. (3) Cooling system. (4) Water transport pipes. Water is used as a neutron shield. (5) CCl_4 pumps. (6) Cerenkov counters. (7) Electronic counters. (8) Antineutrino detectors. (9) Capacitors. (10) (14) (15) Watergates. (11) Laboratory for noble gas analysis. (12) Container of liquid N. (13) Reservoir of CCl_4 with a content of 400 000 liters.

IV.3.3. THE CNO CYCLES

The majority of stars originates from a material (the 'interstellar gas') which is not only composed of H and He, but also of heavy elements like C, N, and O. (These heavy elements have been formed in previous generations of stars and returned to the interstellar matter (see Chapter IX).) C, N, and O are very good catalysts for the transformation of 4H into He. This was first recognized by Bethe* in 1939 who derived the so-called CNO cycle. New measurements of the cross sections of the various implied reactions show that the CNO H burning occurs in three or more cycles as shown in Figure IV.9.

In these cycles, the slowest reactions are $^{14}N(p, \gamma)^{15}O$ and $^{15}N(p, \gamma)^{16}O$, the latter being about one thousand times slower than the rapid one: $^{15}N(p, \alpha)^{12}C$. The (p, α) reactions are the most rapid nuclear reactions and destroy elements like ^{15}N,

* Nobel Prize winner in 1972.

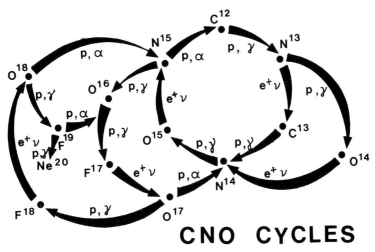

CNO CYCLES

Fig. IV.9. The CNO cycles.

^{17}O, and ^{18}O. As a consequence, the result of the triple cycle is not only the transformation of 4H into He but also the transformation of most of the CNO nuclei into ^{14}N. Table IV.2 gives a comparison between the observed ratios of the abundances of CNO nuclei and the computed ones, at the end of the CNO cycles, in a star in equilibrium conditions (from Trimble, 1975). It is clear that the observed CNO cannot be explained only by this effect. At higher temperatures and in explosive conditions, the CNO cycles lead to a different nucleosynthesis pattern which will be examined in Chapter V.

IV.4. Helium Burning

There is also an important puzzle in the case of He burning: the fusion of 2 He nuclei would give a ^{8}Be nucleus, highly unstable (lifetime of 10^{-16} s). However, if the temperature and density are high enough ($T \simeq 10^8$ K and 10^5 g cm^{-3}), 3 He nuclei can be combined in order to form directly a stable C nucleus. This in fact is a double

TABLE IV.2

Comparison between the observed isotopic ratios of the elements CNO and the computed ones, at the end of the CNO cycles, in a star in equilibrium conditions (after Trimble, 1975).

Ratio	Observed	Calculated
^{12}C/^{13}C	90.	4
^{13}C/^{14}N	0.036	0.018
^{15}N/^{14}N	0.0036	0.0002
^{17}O/^{14}N	0.022	0.02
^{18}O/^{14}N	0.012	very small
^{14}N/^{16}O	0.17	60

nuclear reaction: the reaction ^{4}He + ^{4}He = ^{8}Be produces a very small number of ^{8}Be nuclei (^{8}Be/^{4}He = 10^{-9} at equilibrium). These very few ^{8}Be can in turn combine with other ^{4}He nuclei and give ^{12}C (Figure IV.10). These reactions provide a substantial amount of ^{12}C but only if the cross section of the ^{8}Be + ^{4}He → ^{12}C is enhanced by some resonant effect. Hoyle has suggested in 1955 the existence of an excited state at stellar energies, i.e. 7.6 MeV, in the compound ^{12}C nucleus. This prediction made on astrophysical grounds was subsequently confirmed by laboratory experiments of nuclear physics.

This reaction is not the only one which occurs in the center of stars, at such temperatures (some hundreds of million degrees): the products of the CNO triple cycle are also reacting with He nuclei to give heavier elements:

$$^{12}\text{C} + {}^4\text{He} \rightarrow {}^{16}\text{O} + \gamma$$

$$^{14}\text{N} + {}^4\text{He} \rightarrow {}^{18}\text{F} + \gamma \rightarrow {}^{18}\text{O} + e^- + \nu$$

$$^{16}\text{O} + {}^4\text{He} \rightarrow {}^{20}\text{Ne} + \gamma .$$

These reactions are resonant (i.e. at the considered temperatures their Gamow peak lies in the resonance domain). Their cross section depends on the excited levels of ^{16}O, ^{18}F and ^{20}Ne. In particular, although a tremendous effort has been devoted to its determination, the cross section of the ^{12}C$(\alpha, \gamma)^{16}$O reaction is not precisely known, mainly because this reaction competes with the ^{13}C$(\alpha, n)^{16}$O. At the end of this chain of reactions, there is about 2% of ^{18}O formed. This is enough to induce

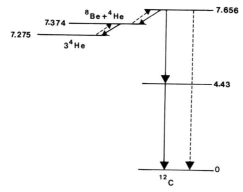

Fig. IV.10. The triple alpha reaction (schematic): The energy diagram of ^{12}C is represented, together with the energies of three ^{4}He and ^{8}Be + ^{4}He when the energy of the fundamental level of ^{12}C is taken equal to zero. The energies are in MeV. It is possible that two ^{4}He nuclei combine and form one ^{8}Be (−0.099 MeV). Although the inverse reaction ^{8}Be → 2 ^{4}He is much more probable, it is then possible that ^{8}Be combines with another ^{4}He and forms a ^{12}C nucleus in the second excited level, with an energy of 7.656 MeV. This excited ^{12}C may decay into three ^{4}He again, or into a ^{12}C in the fundamental level. It may be an indirect decay through the first excited level, with the emission of two photons, or a direct decay with the emission of a pair e^+ e^-. (The direct decay with a photon emission is forbidden by the nuclear selection rules.)

another chain of reactions which is very important since it releases free neutrons:

$$^{18}O \; + \,^4He \rightarrow \,^{21}Ne \; + n$$

$$^{18}O \; + \,^4He \rightarrow \,^{22}Ne \; + \gamma$$

$$^{22}Ne \; + \,^4He \rightarrow \,^{25}Mg + n$$

$$^{22}Ne \; + \,^4He \rightarrow \,^{26}Mg + \gamma$$

$$^{25}Mg + \,^4He \rightarrow \,^{28}Si \; + n$$

$$^{26}Mg + \,^4He \rightarrow \,^{29}Si \; + n \,.$$

As will be seen in Chapter VI, these free neutrons are further absorbed by metals to build up elements heavier than Fe.

IV.5. Hydrostatic C, O, and Si Burning

C burning occurs when the temperature reaches 5 to 8×10^8 K. As for O, its burning occurs at somewhat even higher temperatures ($\sim 10^9$ K). C nuclei react in the following way:

$$^{12}C + \,^{12}C \rightarrow \,^{23}Na \; + p$$

$$\rightarrow \,^{20}Ne \; + \,^4He$$

$$\rightarrow \,^{23}Mg + n \,.$$

The third reaction is endothermic. However, it can occur at high temperatures and becomes an additive neutron source. The reaction

$$^{12}C + \,^{12}C \rightarrow \,^{24}Mg + \gamma$$

is possible but has a negligible rate, in comparison with the others. ^{24}Mg can however be formed by the reaction of He nuclei and protons respectively on ^{20}Ne and ^{23}Na:

$$^{20}Ne \; + \,^4He \rightarrow \,^{24}Mg + \gamma$$

$$^{23}Na + p \rightarrow \,^{24}Mg + \gamma \,.$$

Since the ^{24}Mg nucleus is the most stable in all this series, at the end of the chain of reactions, it becomes the most abundant.

O nuclei undergo the following reactions:

$$^{16}O + \,^{16}O \rightarrow \,^{32}S \; \; + \gamma$$

$$\rightarrow \,^{31}P \; \; + p$$

$$\rightarrow \,^{28}Si \; \; + \,^4He$$

$$\rightarrow \,^{24}Mg + 2 \,^4He$$

$$\rightarrow \,^{31}S \; \; + n$$

^{28}Si is also one of the most stable elements in this chain of reactions, and becomes the most abundant.

These chains of reactions are the last fusion cycles which can occur inside stars. The fusion of two Si nuclei, for example, needs temperatures higher than 4.5×10^9 K. However, at such temperatures, the number of photons liberated by all kinds of reactions so large that they photodisintegrate the nuclei. The less stable nuclei (e.g. the nuclei with odd atomic masses) are destroyed in favor of the most stable ones (e.g., Fe, Si even atomic mass nuclei). This explains why the elements around Fe are more abundant than their neighbors.

Another process which occurs in stars under these conditions is generally referred to as the Si quasi equilibrium. At temperatures of about 3×10^9 K and for densities of about 10^8 g cm^{-3}, He nuclei are liberated by the reaction:

$$\gamma + {}^{28}\text{Si} \rightleftarrows {}^{24}\text{Mg} + {}^4\text{He}.$$

This reaction is very slow due to the large stability of the Si nuclei. However, these He nuclei can in turn be captured by other Si nuclei and lead to S:

$$^{28}\text{Si} + {}^4\text{He} \rightleftarrows {}^{32}\text{S} + \gamma.$$

The He nuclei which have thus disappeared through the second reaction can be provided again by the first one, which leads to a slow shift towards heavier elements; then S itself can fix an ^{4}He and so on. On the other hand, ^{4}He nuclei are also provided by the rapid photodisintegration of ^{24}Mg and of lighter elements:

$$^{24}\text{Mg} + \gamma \rightarrow {}^{20}\text{Ne} + {}^4\text{He}.$$

$$^{20}\text{Ne} + \gamma \rightarrow {}^{16}\text{O} + {}^4\text{He}.$$

Thus heavy elements can be built up by this quasi equilibrium process up to Fe.

IV.6. Conclusion

The nuclear reactions in hydrostatically stable stars do explain the large amount of radiated energy, as well as their evolution from birth to death (see Chapter III). Their stability on the Main Sequence is due to H burning reactions. Afterwards, they evolve towards the giant star (He burning) then towards the supergiant star stage (C burning) and so on. However, these nuclear reactions are not too successful in explaining the observed abundances of elements in the Universe. The light elements i.e., D, Li, Be, and B, are not formed but destroyed by nuclear reactions in stars. He is formed within the stars, but not in a large enough amount to account for its observed abundance. Elements are formed up to the Fe peak (and, as will be seen in Chapter VI, much heavier elements can be also formed in stars by neutron absorption). However, the isotopic patterns observed on stellar surfaces and within the Solar System itself remain unexplained. For example, observed nuclei as ^{25}Mg, ^{26}Mg, ^{29}Si, ^{30}Si and many other ones are not formed in hydrostatically stable stars. It is now believed that He and D (and maybe ^{7}Li?) could have been formed at the origin of the Universe (Chapter VII). ^{6}Li and some ^{7}Li, Be and B are formed in the interstellar matter due to cosmic rays (Chapter VII). Now, the observed pattern of heavy elements is known to be a product of *stellar deaths*. Let us note that many stars die violently in a big explosion: the *supernova* phenomenon. During this explosion many nuclear reactions take place which completely change the element abundances. This will be studied further on in Chapter V.

References

Quoted in the text:

Audouze, J. (ed.): 1977, *CNO Isotopes in Astrophysics*, D. Reidel Publ. Co., Dordrecht, Holland.
Clayton D. D.: 1968, *Principles of Stellar Evolution and Nucleosynthesis*, McGraw-Hill Book Co., New York.
Evans, R. D.: 1955, *The Atomic Nucleus*, McGraw-Hill Book Co., New York.
Trimble, V.: 1975, *Rev. Mod. Phys.* **47**, 877.

Further readings:

Blatt, J. M. and Weisskopf, V. F.: 1952, *Theoretical Nuclear Physics*, John Wiley & Sons, Inc., New York.
Burbidge, E. M., Burbidge, G. R., Fowler, W. A., and Hoyle, F.: 1957, *Rev. Mod. Phys.* **29**, 547.
Fowler, W. A., Caughlan, G. R., and Zimmerman, B. A.: 1967, *Ann. Rev. Astron. Astroph.* **5**, 525.
Fowler, W. A., Caughlan, G. R., and Zimmerman, B. A.: 1975, *Ann. Rev. Astron. Astroph.* **13**, 69.
Reeves, H.: 1964, *Stellar Evolution and Nucleosynthesis*, Gordon and Breach Science Pubs., Inc., New York.

See also: Proceedings of the conferences on neutrinos which appear every two years (e.g. Neutrino '78).

CHAPTER V

EXPLOSIVE NUCLEOSYNTHESIS IN STARS

As seen from Chapter II, the evolution of the stars depends essentially on their masses: low mass stars evolve into white dwarfs at the very end of the planetary nebulae phase (Figure V.1) while high mass ones become neutron stars after having exploded at their supernovae stage (Figure V.2), or black holes.

The limiting mass between white dwarfs and supernovae progenitors is not exactly known because of the uncertainty on the mechanisms of the mass loss processes and the supernovae explosions. However, it is now believed to be between 3 and $8M_\odot$.

During the planetary nebula phase, low mass stars are quietly ejecting their external layers with velocities of about 10 to 30 km s^{-1}. The value of the ejected mass lies between 0.5 and $2M_\odot$. This process lasts for about 10^5 yr. It means roughly that *one* new planetary nebula can be observed per year in *one* galaxy. Its central star starts contracting until the density of its inner zones becomes larger than 10^6 g cm^{-3}. For such large densities, the electrons in the central regions of the star fill up all the lowest energy states; at this point, the electron gas is called 'degenerate'. The pressure of the degenerate electron gas then stabilizes the star which stops its collapse and becomes a white dwarf ($T_{\text{superficial}} \sim 40\,000$ K, $R \sim R_{\text{Earth}}$ i.e. 6×10^8 cm). The whole process is slow and smooth and does not induce any new nuclear reactions. The matter ejected into the interstellar gas has the same composition as the one observed at the end of the red giant phase.

The situation is very different in the case of a supernova explosion. In this case, the temperature becomes so high ($T > 4$ to 5×10^9 K) during a short time scale ($\tau \sim 1$ s) that many nuclear reactions do occur and usually transform the final abundances of the ejected mass. These processes are called 'explosive nucleosynthesis' and are analyzed later in this chapter. It will also be seen that explosive nucleosynthesis can occur during smaller explosions which are not always related to the stellar death (like the novae explosions—see below).

V.1. Supernovae

Supernovae have been discovered and studied since antiquity. The supernova, the remnant of which is the Crab Nebula (Figure V.2), exploded in 1054 (A.D.) and has been extensively observed by Chinese astronomers. There are a few other 'historical' supernovae among which are those discovered by Tycho Brahé (SN 1572) and Kepler (SN 1604).

Supernova explosions are characterized by a sudden extremely strong outburst of energy (10^{51} erg during the whole phenomenon) which provides a maximum

Fig. V.1. A planetary Nebula (NGC 3587) (see also Fig. 1.3). (*Photograph kindly lent by the Observatoire de Haute Provence du Centre National de la Recherche Scientifique, France.*)

luminosity of about $10^9 L_\odot$ (i.e. nearly 1% of the whole galactic luminosity concentrated in a single object!). This maximum luminosity is so large that the Crab Supernova has remained visible even in the daylight during one month. After the explosion, the luminosity decreases with time (see the light curve in Figure V.3). At the time of the explosion, the radius of the star is $> 10^{15}$ cm and the matter is expelled with velocities of $\sim 10\,000$ to $20\,000$ km s^{-1}.

Supernovae are classified in two different types according to the shape of their light curve (although this classification is nowadays sometimes challenged). Both of these types show a very steep decrease (exponential shape) of their light curve. Type I supernovae have a rather regular exponential shaped curve. Their spectrum does not show any H line. The matter which is expelled from these objects is apparently enriched in heavy elements. Type I supernovae are generally found in the old star population regions like the galactic halo or in the elliptical galaxies.

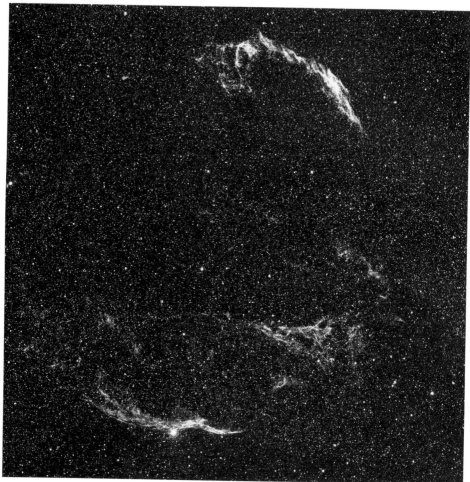

Fig. V.2. A Supernova Remnant (NGC 6960–79–92–95) (see also Fig. 1.4). (*Photograph kindly lent by the Observatoire de Haute Provence du Centre National de la Recherche Scientifique, France.*)

Type II supernovae are associated with young stars: they are only found in the arms of spiral galaxies. Their light curve decreases much more irregularly than that of type I supernovae. He is present in the ejected matter which seems to have normal abundances. Up to now, there is only one case (Cas A) where significant enrichment in O, S, and Ar within the ejected matter has been observed. The Crab Nebula shows an enhancement only for He.

The overall rate of occurrence for supernovae explosions is about one supernova/30 yr per galaxy. The explosion of a supernova does not only lead to a strong outburst of energy but also to the formation of a remnant which remains energetic for a rather long time (10^4 to 10^5 yr). For instance, the Crab Nebula appears as a strong X-ray source. The emitted energy is due to the acceleration of relativistic electrons by magnetic synchrotron radiation. The remaining star becomes a

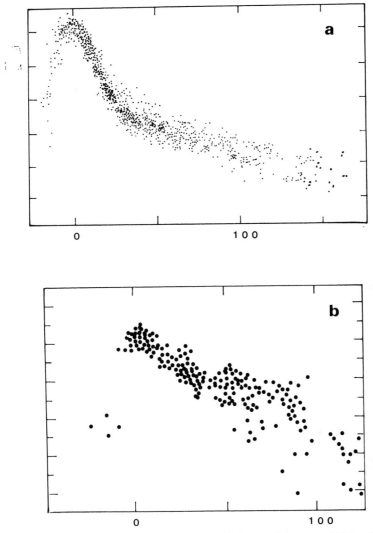

Fig. V.3. (a) Average light curve of type I supernovae. (b) Average light curve of 17 type II supernovae.

neutron star (i.e., a source of the pulsar phenomenon). A pulsar has been associated with a few supernovae remnants like the Crab Nebula.

The mechanism which triggers the supernovae explosions must be able to release a tremendous amount of energy (10^{51} erg is a large energy even on a galactic scale) and also to allow the formation of a remnant. Although many attempts have been made to explain these two facts, only a few hypotheses are able to account for both. We will successively review the mechanisms based on the Fe photodisintegration, the C detonation, the neutrino transport and the rotation energy transfer from the central pulsar.

V.1.1 THE Fe PHOTODISINTEGRATION MECHANISM

This mechanism can occur in the case of stars more massive than $8M_\odot$. In the central regions of such stars, the temperature is $T \geq 5 \times 10^9$ K which allows the formation of Fe. However, at such temperatures, photodisintegration processes can take place (Chapter IV), so that ^{56}Fe can be destroyed:

$$^{56}\text{Fe} + \gamma \rightleftarrows 13\,^4\text{He} + 4\text{n} - Q.$$

Since the ^{56}Fe nucleus is very stable, this reaction is highly endoenergetic and may induce a violent collapse of the central regions of the star. This gravitational collapse is such that the upper layers are precipitated at higher temperatures. Thermonuclear reactions then operate and release a large amount of thermonuclear energy on a very short time scale: the implosion of the Fe core is rapidly followed by the explosion of the matter which has not yet finished its nucleosynthetic evolution. This explosion can explain the supernova phenomenon. In the same time the core of Fe can be neutronized i.e., the protons in the nuclei can capture electrons and be transformed into neutrons. The central part of the supernova then becomes a neutron star ($\rho_c > 10^{13}$ to 10^{14} g cm^{-3}) which can be detected as a pulsar.

V.1.2. THE C DETONATION MECHANISM

The C detonation in principle can take place in stars of mass 4 to $8M_\odot$ which end up their evolution at the stage of C burning. In these stars, the central density is so high that the gas of electrons becomes degenerate, the electrons occupy all the lowest energy states. As said before, in a degenerate gas the pressure is decoupled from the temperature (it depends only on the number of particles). In this case nuclear reactions become explosive: in a normal gas, the pressure acts as a thermostat i.e., an increase of T induces a dilatation which means a pressure decrease which equilibrates the temperature. In contrast, nothing prevents the temperature from increasing in a degenerate gas allowing runaway nuclear reactions (which are very sensitive to the temperature variation).

The explosive burning of C in a degenerate core can trigger a supernova type explosion but in this case a remnant cannot survive. This explosive process also induces a large production of Fe which is in disagreement with the observations. In order to make this C detonation model possible, efforts have been made to find out a way of releasing the energy at the center before it becomes degenerate. Among the proposed mechanisms, we can quote: (a) The effect of the rotation of the central zones: if the nuclear energy is rapidly converted into rotational energy instead of thermal energy, one might avoid the disruption of the central zones. (b) The C burning reactions may generate energetic photons which can be transformed into electron-positron pairs. This last transformation absorbs about 1 MeV per e^+e^- pair formed. (c) The energy dissipation could also be due to the formation of neutrinos according to the so-called Urca process (named after a gambling house in Rio de Janeiro, now closed because of the immoral way in which gamblers were robbed by the tenants!). In this process, neutrinos are produced by a sequence of electronic captures and

β disintegration:

$$(A, Z) + \bar{e} \rightarrow (A, Z - 1) + \nu$$
$$(A, Z - 1) \rightarrow (A, Z) + e^{+} + \nu.$$

Neutrinos have a very small cross section of interaction with the matter and then can leave the central regions with their own energy. Furthermore, it appears that the Urca process is especially favored by nuclei such as ^{23}Na which is copiously formed during the C burning process. In principle, this process can be invoked to avoid the disruption of the central regions of the stars. Unfortunately, from recent calculations taking into account the effects of the convection and the variations of the entropy of the stellar matter submitted to the Urca process, it seems impossible to avoid the total disruption of the central regions of the 4 to $8M_\odot$ stars. Therefore, the fate of these stars is not yet accurately known.

V.1.3. THE NEUTRINO TRANSPORT MECHANISM

The fusion reactions generate energy which is transported not only by the photons but also by neutrinos which are formed in greater quantities as the fusion reactions proceed to heavier nuclei. The neutrinos have a very small interaction cross section with the matter ($\sim 10^{44}$ cm^{-2}, i.e. $\sim 10^{-17}$ times a typical nuclear cross section). This cross section increases with the square of their energy. Colgate and White (1966) have proposed a mechanism in which the energetic neutrinos which are produced in the central regions of the stars can be absorbed by the external layers. These layers in which the neutrinos deposit their energy are heated and can subsequently ignite and explode.

This model is not free from difficulties. For instance, the fusion reactions create not only electronic neutrinos which can eventually be absorbed by the outer stellar zones but also (in comparable amounts) muonic neutrinos which have a much smaller interaction cross section with this gas. However, this mechanism seems to find new support from the recent works on weak interactions, in particular with the discovery of the neutral currents and the possibility for the family of the neutrinos to have other members which can more strongly interact with the hadronic matter. Moreover, one has recently noticed that the neutrino interaction cross section is not only proportional to E^2 but also possibly to A^2 (where A is the mean atomic mass of the nuclei of the gas)$\cdot$ This A^2 effect takes place only in a non-degenerate gas and may favor the neutrino absorption by non-degenerate external layers (Figure V.4). New calculations of cross sections of neutrino/matter interaction make this model quite plausible.

V.1.4. DECELERATION OF THE CENTRAL PULSAR

The previous models assumed that the pulsar was formed during or after the supernova explosion. Another type of model supposes that the supernova explosion is a consequence of the formation of the pulsar. According to these last models, the central zones are first neutronized and only then do they form a neutron star or pulsar (Figure V.5). The pulsar may transfer its rotation energy to the external

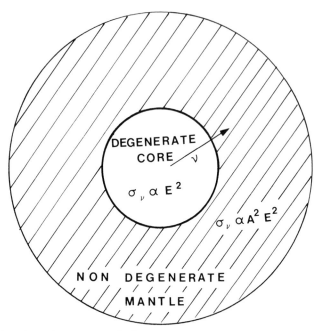

Fig. V.4. Neutrino absorption processes: in the core the matter is electron degenerate and the neutrino absorption cross sections depend only on the energy (in E^2). In the non-degenerate mantle the neutrino absorption cross sections increase as $E^2 A^2$ where A is the average atomic mass of the mantle material. This figure explains why the neutrinos which are produced in the core can be absorbed efficiently in the mantle region.

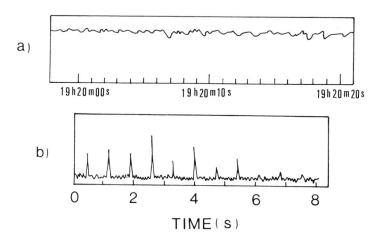

Fig. V.5. (a) The first chart record of individual pulses from a pulsar, PSR 1919 + 21, recorded on November 28, 1967. Increasing intensity is downwards on the chart. *(From Hewish, 1975, Copyright American Association for the Advancement of Science.)* (b) Chart record of individual pulses from one of the first pulsars discovered, PSR 0329 + 54. They were recorded at a frequency of 410 MHz. The pulses occur at regular intervals of about 0.714 s.

zones through the hydromagnetic waves created by the torque between the rotation and the magnetic field. Current estimates for the rotation energy (the pulsars have periods between 10^{-3} and 1 s), for the magnetic fields ($B \sim 10^{12}$G) and for the characteristic time scale of pulsar braking lead to energy values quite sufficient to explain the supernova explosions.

V.2. Other Explosive Objects or Explosive Stages

The supernovae are not the only explosive objects. Important outbursts are also taking place during the He flashes which occur within the red giant phase of the stellar evolution. Novae outbursts are less spectacular than supernovae explosions but are also interesting as far as the nucleosynthesis of rare species such as ^{13}C, ^{15}N and ^{17}O is concerned. Finally, theorists have advocated that the energetic outbursts taking place in active nuclei of galaxies such as the Seyfert galaxies or in quasi-stellar objects (quasars), might be explained by the explosion of very massive stars ($M > 10^4 \, M_\odot$) which also could have some nucleosynthetic consequences.

V.2.1. THE HELIUM FLASHES

During the red giant phase, the electrons of the He burning zone become degenerate. The degeneracy phenomenon has the already mentioned consequence of decoupling the temperature and the density. The He burning zone can therefore experience partial runaways which are called flashes. These flashes have many nucleosynthetic outcomes because they induce a partial mixing of the He and H burning zones: this mixing can be very important in the nucleosynthesis of various elements such as ^{7}Li, ^{13}C and the s process elements. As it will be shown in Chapter VII, ^{7}Li can be produced by the reaction ^{3}He + ^{4}He → ^{7}Be ($+$ e$^-$) → ^{7}Li. ^{3}He comes from the H burning zone and the reaction is favored by the mixing of ^{3}He with ^{4}He coming from the degenerate zone. If ^{7}Be which is then produced reaches low temperature zones, it can be transformed into ^{7}Li by electron capture. There are a few red giants which are indeed overabundant in ^{7}Li (Li/H $\sim 10^{-7}$ i.e., 100 times the interstellar medium value).

As for ^{13}C, it can be formed during these flashes: the mixing is such that the amount of H mixed with the He zone is smaller than the amount of ^{12}C present in the He burning zone. In this case, the CNO cycle is incomplete and cannot proceed further than the reaction ^{12}C(p, γ) ^{13}N(β^+) ^{13}C. In fact, significant enrichments in ^{13}C are observed in red giants.

As will be seen in Chapter VI which is devoted to the nucleosynthesis of elements with an atomic mass heavier than Fe, the s-process elements are formed by *slow* processes of neutron absorption by the Fe peak nuclei. This process takes place indeed during the red giant phase of the stellar evolution: unstable nuclei so formed, like Tc ($t_{1/2} \sim 2 \times 10^5$ yr) and possibly Pm* ($t_{1/2} \sim 25$ yr), have been detected by their characteristic lines in the optical spectrum of cool red giants. Furthermore, there exists at least two very evolved stars, FG Sagittae (studied by the astronomers

---

* Although in this case, the detection of the Pm line is not proved without ambiguity.

of Lick) and CI Cygni (studied by the astronomers of the observatoire de Haute Provence), for which one has discovered that the abundances of the s-process elements like Ba, Y, or Zr increase by factors as large as 3 to 5 on times scales as short as several months or one year (which is extremely rapid if compared to the stellar evolution time scales). The formation of these s process elements can take place during or after the He flashes occurring during the red giant phases: the mixing of H and He zones can trigger thermonuclear reactions releasing neutrons. One of the most important reactions of this type is the following: $^{13}C + {}^4He \rightarrow {}^{16}O + n$, where the H rich zone is highly enriched in ^{13}C via the previously described process.

These three examples show the importance of runaways or instabilities, during the normal course of stellar evolution, on the nucleosynthesis.

V.2.2. THE NOVAE OUTBURSTS

Novae are explosive objects which are quite different from supernovae in many respects: during their outbursts they emit an energy of 10^{42} to 10^{44} erg which is basically $10^{8 \pm 1}$ times smaller than the energy released by the supernova explosions. In fact, the novae are a collection of somewhat different objects: some of them are called dwarf novae and release energies much lower than the values quoted above. On the other hand the nova outburst which occurred in the Cygnus constellation on August 1975 released an energy as large as 10^{46} to 10^{47} erg which makes this specific object intermediary between ordinary novae and supernovae. One can observe a few novae explosions (up to 10) within a given galaxy in one year, which means that the rate of novae outbursts is about 10^3 times larger than the supernovae one. Therefore, energetically speaking, the novae outbursts appear less important than the supernovae explosions. However, the formation of rather rare nuclear species like ^{13}C, ^{15}N, or ^{17}O can be attributed to novae through the explosive CNO cycles which will be reviewed later on. During the nova outbursts which last for times comparable to those of the supernovae explosions (Figure V.6) the object rejects about 10^{-4} to 10^{-3} $M_\odot$ with velocities of ~ 1000 km s^{-1}. Furthermore, it seems that the ejected matter or the matter at the surface of the nova is enriched in C, N and O, and possibly in ^{13}C relative to ^{12}C, and ^{15}N relative to ^{14}N. Novae are generally recurrent objects: they can experience several outbursts. Contrary to supernovae which are almost entirely destroyed by the explosion, the nova outburst affects only its surface. Two Soviet astronomers, Kukarkin and Paranego, discovered an important relation between the periodicity of two outbursts and the energy released during these outbursts: the larger the amount of energy released, the longer the time between two runaways. The Kukarkin-Paranego relation can be understood in the framework of the outburst theory, making use of the fact that novae always belong to double star systems. In double star systems, one can define a surface of points equally attracted by each of the members of the binary system (Figure V.7), this surface is called the Roche lobe. The prenova is generally a white dwarf which is supposed to be CNO enriched. Let us now assume that its companion becomes a red giant with a radius theoretically larger than the Roche lobe. The external matter of the companion, mainly made out of H and He, is then attracted by the prenova and violently accreted by it. This material heated at temperatures

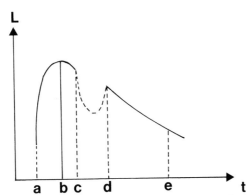

Fig. V.6. Typical Nova light curve (in optical wavelengths). The average increase of luminosity ranges from factors 10^3 to 10^6. The maximum lasts a few weeks. The cd period is called the transition period during which the nova often has a maximum luminosity in the IR (dust formation?). During the last period the luminosity is smoothly decreasing logarithmically. The shapes of these light curves vary much from one object to the other because they depend much on the characteristics of the accretion mechanism.

$T > 10^8$ K triggers the hot CNO burning cycle. The important energy released by this hot cycle induces the outburst of the layer corresponding to these reactions. This rather straightforward process does not only account for the nova outburst but also provides a simple explanation of the Kukarkin-Paranego correlation. If the period is long enough, the nova has time to accrete more material; therefore the released energy can be greater and the outburst more violent. This model of nova outburst is rather adequate to explain the main observed features of these events; it relates this outburst to some nuclear (or nucleosynthetic) cause, as in the case of supernovae.

V.2.3. EXPLOSIONS OF SUPERMASSIVE STARS

The supermassive stars have been invented by theoreticians. They are stellar objects of mass $10^4 < M < 10^6 \, M_\odot$ which can evolve according to three possible schemes depending on their mass and their content in heavy elements Z (Figure V.8).

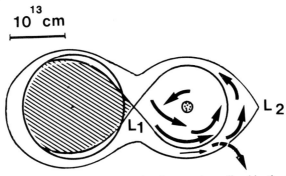

Fig. V.7. Binary star model of nova outburst showing the accretion suffered by the prenova white dwarf. The matter falling on its surface comes from its companion which entirely fills up its Roche lobe.

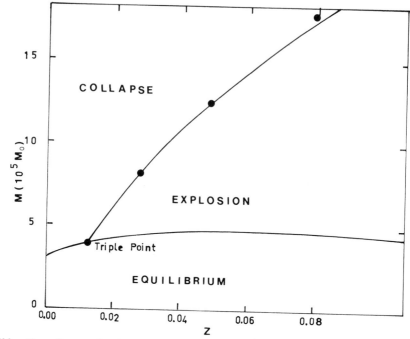

Fig. V.8.   Phase diagram showing transition masses as a function of heavy element content for initial H abundance $X_0 = 0.70$. In the collapse region, the supermassive stars collapse and become a huge black hole. In the explosion regions the supermassive stars first collapse and then their central region suffers hot CNO nucleosynthesis processes which halt the collapse and make the supermassive stars explode.

(1) The very massive stars with a low Z contract violently and become black-holes.

(2) The less massive objects $M < 5 \times 10^5 \, M_\odot$ evolve like ordinary stars in several successive phases of collapse and nuclear burning.

(3) The massive objects $M > 5 \times 10^5 \, M_\odot$ with a significant Z abundance first contract because of the gravitation. Then, the nuclear reactions take place and stop the contraction which leads to an explosion similar to those discussed above. This last case is the most interesting one to explain the nuclei of active galaxies and some of the observed features of the quasars. They also have some nucleo-synthetic activity during the explosion.

V.3. The Explosive Nucleosynthesis

In all the events which have been described above, the explosion is generally trig-gered by some nucleosynthetic event (it is more often some thermonuclear reaction) which is able to release a large amount of energy and to lead to the formation of various elements such as ^{13}C, ^{15}N in the case of novae outbursts or to the formation of the Fe peak nuclei during (or just before) the explosion of the supernova. The nuclear reactions which occur and cause these explosions, take place on shorter time scales (the time for the explosion is generally from a few seconds to a few hours), at much higher temperatures and/or in denser regions, than in hydrostatically stable stars. The temperature is indeed the most important factor governing the nucleosynthesis which occurs during stellar explosions, since the reaction rates strongly depend on it.

In any problem of nucleosynthesis and in particular in a problem of explosive nucleosynthesis, one has to follow the variation with time of the abundances of the nuclear species involved in the nuclear process. The problem is then to solve a set of differential equations. It is not the purpose of this review to describe the specific ways to solve such systems. This problem can be quite complicated in the most general case where the nuclear species i can be both destroyed by interaction with the species j and formed in the reaction between two other particles k and l. One must introduce in the computations the abundances of all species, and the reaction rates $\langle \sigma v \rangle_{i,j}$ for the destruction of i by reaction with j and $\langle \sigma v \rangle_{k,l}$ for the formation of i reactions between k and l. The method used in attempting to solve such problems is to linearize these equations by dividing the time range into very short time intervals. The problem is then to solve big sets of linear equations. Hopefully, modern computers make this type of job quite easy to solve.

The expressions for the nuclear reaction rates are the same as those derived in Chapter IV. Since the time scales of the explosions are short, the dependence on the temperature and the density with time during the explosion is of major importance in the computations of the final abundances. In fact, there are three different problems which should be solved simultaneously: one should start from a model of stellar evolution which mainly depends on the thermodynamics of the gas (the equation of state of the gas) and on the way by which the energy is transported within the different zones. Explosions involve a hydrodynamical treatment of the explosive events i.e. the way according to which shock waves, combustion and detonation waves, are propagating through the zones with time, or in other words the dependence of the density and temperature parameters with time. When the stellar evolution and the hydrodynamics of the gas are properly described, one can try to solve the equations describing the nucleosynthesis able to take place in the gas. Actually, these three problems are not really separated but are interacting with one another: the nuclear reactions release an energy which plays a major role in the stellar evolution and especially in the hydrodynamics of the explosion. The hydrodynamics govern the time scales of the temperature and the density variations and furthermore the way by which the nucleosynthesis proceeds in a given zone (Figure V.9). This type of problem requires difficult computing and only very few treatments of that kind have been made.

Nevertheless, one can already have some rough insight on the possible types of nucleosynthetic processes which can take place in an explosion and on the output of such events by making the following simplifying assumptions:

(1) The most drastic hypothesis is to assume that the nuclear species which participate in the explosive process quickly reach an equilibrium; in this case, their abundances are simply inversely proportional to the rate of nuclear reactions destroying them. This very crude simplification still provides a useful indication on the nucleo-

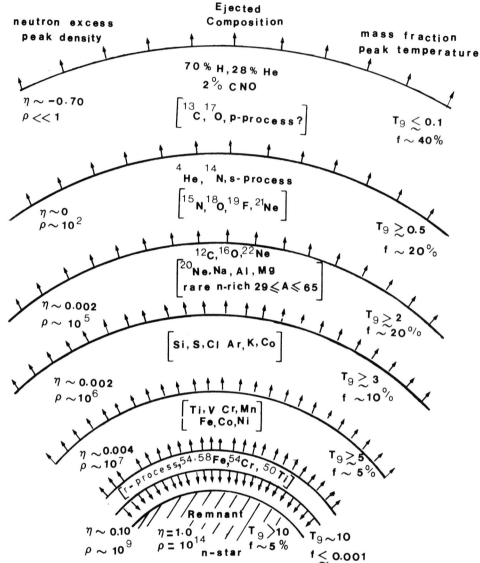

Fig. V.9. Theoretical onion skin model for a large mass star ($M > 10M_\odot$) before the supernova ignition. Each layer is defined by its density ρ (in g cm^{-3}), its temperature (T_9 in units of 10^9 K), its relative mass f, its neutron excess η (see definition, p. 75) and its chemical composition.

them. This very crude simplification still provides a useful indication on the nucleo-synthetic effects that one can expect for a given set of temperature and density conditions. It is only applicable to the case where few reactions are being considered.

(2) When the number of reactions become important, namely when a given nucleus can be destroyed or formed by more than one reaction, one cannot assume

anymore that equilibria are taking place for the whole process. A less stringent simplification is then to consider that, during the outburst, temperature and density still remain constant but that the nuclear species are no longer in equilibrium. Although this treatment is still oversimplified (actually the thermodynamical parameters are drastically evolving during the shock), it provides interesting indications on the final results because of the maximum temperature caused by the shock which is the most critical parameter in these calculations.

A better treatment is to assume that the temperature and the density evolve according to a theoretical 'profile' for which the characteristic time scale is the so-called free fall time scale:

$$t_c \sim \frac{446}{\sqrt{\rho_0}} \text{ (s)}$$

where the initial density ρ_0 is expressed in g cm^{-3}. The reader can derive this expression for t_c by simply using the Newton law for stars, all pressure forces being neglected

$$f = m\frac{d^2r}{dt^2} = -\frac{GMm}{r^2}.$$

By multiplying both sides of the equation by dr/dt and integrating, one obtains:

$$\frac{1}{2}\left(\frac{dr}{dt}\right)^2 = \frac{GM}{r} = \frac{4\pi r^2 \rho G}{3}$$

where the integration constant is neglected, so that

$$\frac{1}{r}\frac{dr}{dt} = \left(\frac{8\pi G\rho}{3}\right)^{1/2} \quad \text{and} \quad \frac{1}{\rho}\frac{d\rho}{dt} = (24\pi G\rho)^{1/2},$$

the free fall time scale is

$$t_c = \left(\frac{1}{\rho_0}\frac{d\rho_0}{dt}\right)^{-1} = \frac{1}{\sqrt{24\,\pi G\rho_0}} = \frac{446}{\sqrt{\rho_0}}.$$

In many explosive situations the initial density is $\geq 10^4$ g cm^{-3} which means that the characteristic time scale can be as low as a few seconds.

If one makes the further assumption that the shock wave propagates adiabatically in the material, the evolution of the temperature and of the density is simply given by:

$$T = T_0 \exp\left(\frac{-t}{3t_c}\right) \quad \text{and} \quad \rho = \rho_0 \exp\left(\frac{-t}{t_c}\right).$$

These exponential profiles were currently used in the classical calculations of explosive nucleosynthesis. A proper treatment of a stellar explosion and its nucleosynthesis should include:

(1) A complete stellar model providing the thermodynamics of the stellar matter.

(2) A hydrodynamical code following the propagation of shock waves or detonation waves (shock waves triggered by nuclear or chemical reactions).

(3) A nucleosynthetic code allowing the determination of the abundance evolution and providing an estimate of the energy released by fusion reactions.

Up to now, the only case where these complete calculations have been performed and published are those concerning the novae outbursts. Complete calculations of supernovae explosions are presently undertaken by different groups and should be completed in the near future.

V.4. The Main Results of the Explosive Nucleosynthesis

In this section, we describe the explosive nucleosynthesis occurring in the H and He rich zones of novae and supernovae as well as the explosive processes occurring in dense regions of a supernova, C burning, O burning and Si burning. A large fraction of the heavy elements (the r and p process elements) originates in explosive processes too. Their nucleosynthesis is analyzed in Chapter VI. The explosive nucleosynthesis which occurs in the primordial phases of the Universe /Big Bang nucleosynthesis) will be reviewed in Chapter VII.

V.4.1. EXPLOSIVE BURNING IN H AND He BURNING ZONES

In these explosive events, the fuel is H and/or He but the elements which suffer the main nuclear transformations are the $A \geq 12$ elements (mainly the CNO isotopes): the so-called 'hot CNO cycle' or 'explosive CNO cycle' which can take place in the external layers of novae and supernovae.

The temperature range for the explosive CNO burning to take place is $10^8 < T < 10^9$ K. At such temperatures, the time scales of the nuclear fusion reactions occur with a rate much faster than for the beta decay reactions and influence the nuclear by-products (Figure V.10). In the 'cold' or 'classical' CNO cycle, the slowest

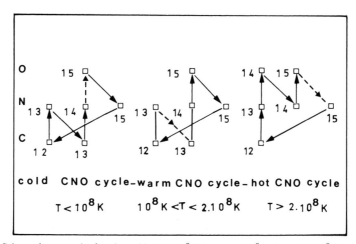

Fig. V.10. Schematic networks for the cold ($T < 10^8$ K), warm ($10^8 < T < 2 \times 10^8$ K) and hot ($T > 2 \times 10^8$ K) CNO cycle. The bottle-neck (slowest reaction) is indicated with a dashed line for these three temperature conditions.

reaction is the ^{14}N(p, γ)^{15}O reaction; as a result the CNO cycle processing in the core of Main Sequence stars or in the H rich zones of red giants does not only lead to the transformation of H into He but also to the transformation of C and O into ^{14}N. Furthermore, since the most rapid nuclear reaction is ^{15}N(p, α)^{12}C, the ^{15}N/^{14}N ratio at the end of this CNO cycle is only 4×10^{-5}, about 100 times smaller than the ^{15}N/^{14}N ratio observed in the solar system.

At temperatures $T > 10^8$ K, the slowest reaction are:

(1) ^{13}N(β^+)^{13}C for $10^8 < T < 2 \times 10^8$ K which favors the formation of ^{13}C in the so-called 'warm' CNO cycle (Figure V.10); the main by-product of this process is then ^{13}C.

(2) At higher temperatures ^{13}N can itself be easily destroyed by ^{13}N(p, γ)^{14}O. The slowest reactions are ^{14}O(β^+)^{14}N and especially ^{15}O(β^+)^{15}N. In some cases the resulting ^{15}N/^{14}N ratio can be larger than 1.3. For temperatures higher than 10^9 K the nucleosynthesis proceeds very quickly, rapidly transforming (in less than a few seconds) all the CNO material into heavier elements. Then, as it has been stated above, the temperature is the main parameter governing the explosive or hot process. These processes are likely to take place in novae outbursts which could be the sites for the formation of ^{13}C and ^{15}N.

If supermassive stars exist, their explosion can be triggered by the hot CNO cycle and lead to interesting overabundances of rare nuclear species such as ^{3}He, ^{7}Li, ^{13}C, ^{15}N, and ^{17}O.

Another important parameter is obviously the composition of the gas where the explosion takes place. In H rich zones, the main products are ^{13}C first and then ^{15}N and ^{17}O, at respectively lower and higher temperatures. In such environments, it is almost impossible to account for the formation of ^{18}O and ^{19}F which have unstable parents (^{18}Ne and ^{19}Ne) with a too short lifetime.

In zones where He is the most abundant component, ^{13}C and ^{17}O are no longer overproduced while ^{15}N, ^{18}O, ^{19}F, and ^{21}Ne can be favored by such processes.

An approximate combination of the explosive H and He burning might provide a reasonable explanation for the formation of relatively rare nuclear species such as ^{13}C, ^{15}N, ^{17}O, ^{18}O, ^{19}F, and ^{21}Ne. Furthermore, in such environments, elements like ^{3}He, ^{7}Li, or ^{11}B (to a lesser extent) can be easily produced from the D(p, γ)^{3}He, ^{3}He(α, γ)^{7}Be(e$^-$)^{7}Li, and ^{7}Be(α, γ)^{11}C(β^+)^{11}B reactions.

V.4.2. Explosive Nucleosynthesis in C, O, and Si Burning Zones

These processes are interesting because they produce the rare isotopes observed in the Universe (Chapter III) but not synthetized in hydrostatically stable stars (Chapter IV): ^{24}Mg, ^{25}Mg, ^{29}Si, ^{30}Si...

The explosive C burning takes place at temperatures $T \sim 2 \times 10^9$ K, densities $\rho \sim 10^5$ g cm^{-3}. The initial composition is also a very important parameter in the evolution of the nucleosynthesis. It is currently parametrized by the so-called neutron enrichment factor η ($\eta = (N - Z)/(N + Z)$ where N is the average number of neutrons and Z the average number of protons). The solar η is $\sim 2 \times 10^{-3}$. With that type of neutron enrichment factor obtained by assuming that the initial composition is 49% of ^{12}C and ^{16}O, respectively, and 2% of ^{22}Ne (coming from the transformation of ^{14}N by He burning: $^4N(\alpha, \gamma)^{18}F(\beta^+)^{18}O(\alpha, \gamma)^{22}$Ne), the elements which are likely to be produced are ^{20}Ne, ^{23}Na, ^{24,25,26}Mg, ^{27}Al, ^{29}Si and ^{30}Si (Figure V.11). As it can be seen from Figure V.11(b) a change in the η factor modifies significantly the output of the explosive burning. Similarly, if the initial temperature rises up to $T = 2.2 \times 10^9$ K (all the other parameters remaining the same), the final by-products are ^{28}Si, ^{31}P, ^{32,33,34,36}S, and ^{35}Cl. One sees that an increase of T by 10% is sufficient to entirely modify the final output of the nucleosynthesis.

Let us mention finally that it is sufficient to process 3% of the matter of the solar system to account for the observed abundances in the $20 < A < 28$ mass range.

The explosive O burning is supposed to take place in O rich zones with the same type of neutron enrichment but at $T = 2.6 \times 10^9$ K and $\rho \sim 2 \times 10^5$ g cm^{-3}. In these conditions, the elements which are likely to be produced are ^{28}Si, ^{32}S, ^{34}S, ^{35}Cl, ^{36}Ar, ^{38}Ar, ^{40}Ca and ^{46}Ti (Figure V.12).

It has been also realized that the explosive O burning leads to similar results as the Si quasi equilibrium (Chapter IV).

The explosive Si burning occurs at temperatures of $\sim 4.7 \times 10^9$ K and densities $\sim 2 \times 10^7$ g cm^{-3} with a neutron enrichment factor $\eta \sim 2 \times 10^{-3}$. In the present published calculations, the cooling time scale has been taken much shorter than the free time scale (generally four to five times shorter). Elements such as ^{32}S, ^{36}Ar, ^{40}Ca, ^{52}Cr and ^{54}Fe can be favored by such nucleosynthetic processes but elements such as ^{42}Ca, ^{54}Cr, ^{56}Fe, ^{58}Ni, ^{58}Fe and ^{54}Mn are overproduced; therefore such explosive processes are not very interesting.

The statistical equilibria due to the photodisintegration of Fe have already been examined: the competition between the fusion and the photodisintegration reactions leads to different results according to the maximum temperature which is reached in the center of the supernova, the rate of cooling, and also to the enrichment into neutrons. For instance, if the enrichment into neutrons is weak, the dominant element is ^{56}Ni, while it is ^{56}Fe if this enrichment is important. For a large T_{max} and a cooling time scale equal to the free fall time scale, the particles which are released by the photodisintegration reactions can be absorbed subsequently and the synthesis of elements like ^{52,53}Cr, ^{54}Mn, and ^{54,56,57}Fe can occur. If the maximum temperature is not as large (5×10^9 K instead of 8×10^9 K) and the cooling much more rapid, elements like ^{54}Fe, ^{56}Fe, ^{59}Co and the Ni isotopes can be formed.

To conclude this survey of the explosive nucleosynthesis, one should notice that explosions often take place in the observable Universe; not only do stars explode but possibly larger bodies like supermassive stars, galactic centers, etc. .... Explosions are triggered by nuclear reactions which release in very short time scales tremendous amounts of energy and can produce various different elements, especially the rarest nuclear species which are not explained by the current 'quiet' nucleosynthesis occurring during the normal course of the stellar evolution (Chapter IV). However, the calculations so far reported, which have been quite successful to reproduce the Solar System abundances, still suffer from three types of difficulties:

(1) Many different parameters (the initial temperature, the initial density, their evolution with time, the initial composition and the relative masses of the different zones) are used to account for the abundance distribution of a rather specialized

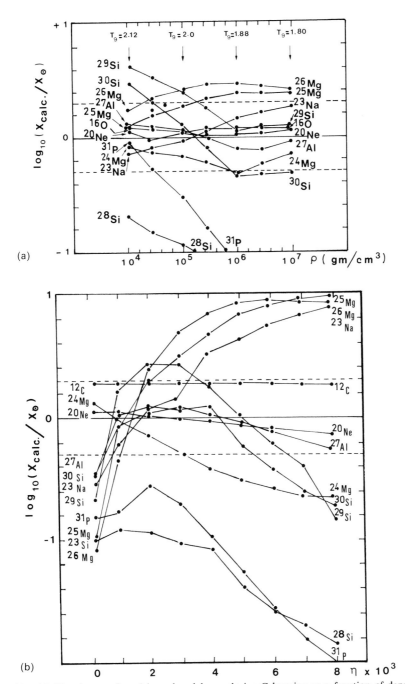

Fig. V.11. (a) Abundances of nuclei produced by explosive C burning as a function of density while varying the temperature in order to burn the same amount of C in each case. (b) Abundances of nuclei produced by explosive C burning as a function of the neutron excess η. One sees that the best fit is reached for $10^{-3} < \eta < 2 \times 10^{-3}$.

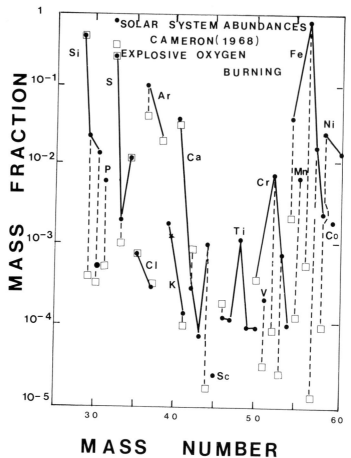

MASS NUMBER

Fig. V.12   The abundance pattern achieved in explosive O burning compared to solar abundances (normalization at ^{28}Si).

and defined amount of matter: the Solar System. Therefore, it is presumptuous to claim that explosive nucleosynthesis explains the solar system abundances when it depends on so many free parameters.

(2) One has not yet solved the full problem: combining the hydrodynamical and thermodynamical study of an explosion occurring in reasonable presupernova models, with the treatment of the nuclear evolution of the exploding gas.

(3) The material forming the solar system does not have a homogeneous composition: recent determinations of isotopic ratios conducted on mineral separates of some carbonaceous chondrites show various and important anomalies in these ratios (this is the case, for instance, for O, Ne, Ca, and Nd). So the solar system composition appears to be a mixture of different components which have different nucleosynthetic backgrounds (see Chapter VIII). It then becomes difficult to define a 'canonical' explosive nucleosynthesis which could explain all of the standard abundances.

References

Quoted in the text:

Audouze, J. and Lazareff, B.: 1976, in M. Friedjung (ed.), *Novae and Related Stars*, D. Reidel Publ. Co., Dordrecht, Holland.
Colgate, S. A. and White, R. H.: 1966, *Astrophys. J.* **143**, 626.
Fricke, K. J.: 1973, *Astrophys. J.* **183**, 948.
Manchester, R. N. and Taylor, J. H.: 1977, *Pulsars*, W. H. Freeman & Co., San Francisco.
Pardo, R. C., Couch, R. G., and Arnett, W. D.: 1974, *Astrophys. J.* **191**, 714.
Rosino, L.: 1977, in D. N. Schramm (ed.), *Supernovae*, D. Reidel Publ. Co., Dordrecht, Holland.
Truran, J.: 1977, in D. N. Schramm (ed.), *Supernovae*, D. Reidel Publ. Co., Dordrecht, Holland.
Woosley, S. E.: 1974, quoted in V. Trimble: *Rev. Mod. Phys.* **47**, 877.

Further readings:

Arnett, W. D.: 1973, *Ann. Rev. Astron. Astrophys.* **11**, 73.
Audouze, J. (ed.): 1976, *CNO Isotopes in Astrophysics*, D. Reidel Publ. Co., Dordrecht, Holland.
Friedjung, M. (ed.): 1976, *Novae and Related Stars*, D. Reidel Publ. Co., Dordrecht, Holland.
Gallagher, J. S. and Starrfield, S.: 1978, *Ann. Rev. Astron. Astrophys.* **16**, 171.
Schramm, D. N. (ed.): 1977, *Supernovae*, D. Reidel Publ. Co., Dordrecht, Holland.
Schramm, D. N. and Arnett, W. D.: 1973, *Explosive Nucleosynthesis*, University of Texas Press, Austin.
Tayler, R. J. (ed.): 1974, 'Late Stage of Stellar Evolution', *IAU Symp.* **66**, D. Reidel Publ. Co., Dordrecht, Holland.
Truran, J. W.: 1973, *Space Sci. Rev.* **15**, 23.

CHAPTER VI

FORMATION OF THE HEAVY ELEMENTS: *s*, *r*, and *p* PROCESSES

As has been seen in the previous chapters, the formation of the elements from C to the Fe peak ($\rightarrow {}^{64}$Ni) is well explained by thermonuclear fusion reactions and quasi-equilibrium processes occurring during either the quiet phases of the stellar evolution (Chapter IV) or the explosive phases (Chapter V). However, light elements ($A < 12$) and elements with an atomic mass $A > 65$ are not significantly synthetized by these processes. Light elements such as Li, Be, B are destroyed more than produced by thermonuclear reactions while the amount of He produced by stars is not sufficient to account for its observed abundance. The problem of their nucleosynthesis is analyzed in Chapter VII.

The situation is quite different for heavy elements: when the atomic mass increases, the Coulomb repulsion between nuclei becomes larger because of the increase of the atomic number: thermonuclear reactions between nuclei with $A > 64$ can occur only at very high temperature $T > 5$ to 6×10^9 K as has been seen in Chapter IV. At such temperatures the photodisintegration processes occur more easily than the fusion processes themselves. This is the reason why fusion processes cannot explain the formation of elements heavier than those of the Fe peak.

The limitation due to Coulomb effects on the fusion reactions between charged particles does not apply to the absorption of neutrons which have no Coulomb barrier to cross before their absorption. Furthermore, neutron absorption reactions have a cross section increasing with the decreasing energy $[\sigma(n, \gamma) \propto (1/v)]$. This is why the formation of most of the heavy elements can be explained by neutron captures. Two very different processes of neutron capture take place and can be distinguished according to the characteristics of the neutron fluxes and of the astrophysical sites. *The s process* (reactions with *slow* fluxes of neutrons) takes place within stable stars during the Red Giant phase when neutrons are released during the normal sequence of thermonuclear reactions (Chapter IV). The nuclear reactions which are the most important neutron sources for the *s* process are in particular ^{13}C$(\alpha, n){}^{16}$O or ^{22}Ne$(\alpha, n){}^{25}$Mg (see below).

The r process (reactions with *rapid* fluxes of neutrons) occurs under explosive conditions when large neutron fluxes can be copiously released during short time scales.

As we shall see further on in this chapter, there are stable heavy nuclei which cannot be formed by neutron captures: they are also called by-passed or *p* process nuclei. Other processes have to be invoked to explain their formation such as proton captures or (γ, n) reactions (photon absorption followed by the release of a neutron) occurring in explosive events, or possibly spallation reactions, or weak interactions . . .

VI.1. Abundances of the Heavy Elements – Processes of Neutron Capture

The Standard Abundance Distribution was discussed and displayed in Chapter III, Figure III.5. The distribution of heavy element abundances ($A > 64$) is represented by two nearly parallel curves. Both of them decrease slowly and rather smoothly with the atomic mass (Figure VI.1). The upper curve represents the abundances of the neutron rich isotopes while the lower curve corresponds to those of the proton rich (by-passed) isotopes. Both curves show characteristic peaks corresponding to particularly stable isotopes as we shall see below.

Stable nuclei can be represented in a diagram (Figure VI.2) which is currently called 'Chart of the nuclides' where the number N of neutrons contained in the nucleus of these nuclides is plotted on the abcissa and the number Z of protons on the ordinate. On this diagram, the isotopes of a given element (same name and atomic number Z) are on a horizontal line. The isobars (same atomic mass $A = N + Z$)

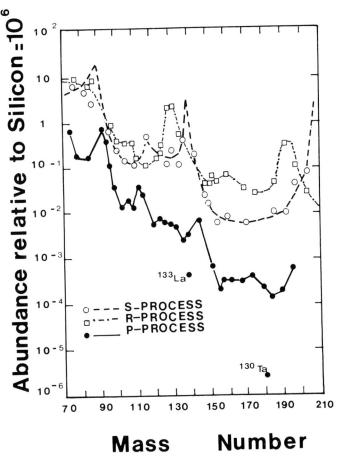

Fig. VI.1.   Abundances of s, r and p process nuclei as a function of atomic mass.

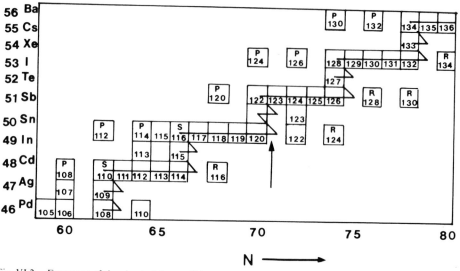

Fig. VI.2.   Fragment of the chart of the nuclides on which the number of neutrons N in a given nucleus is on the abscissa while the number of protons (atomic number Z) is on the ordinate. The isobars (nuclides with the same atomic mass A) are on diagonals. The nucleosynthetic path of the s process is presented on this figure. For instance, ^{116}Cd is a pure r nucleus while ^{116}Sn which is shielded from the r process by ^{116}Cd which is a pure s nucleus. On the left of the diagram, elements such as ^{108}Cd or ^{112}Sn which are 'by-passed' by the neutron absorption path are the p process elements.

are on diagonals. On this diagram, the stable nuclides are referred to as s, r or p according to their formation processes which are discussed below.

The nuclear species with even atomic numbers (Z) have several stable isotopes: this is the case for instance for Sn $(Z = 50)$ which has ten different natural isotopes. One to three different nuclear species may correspond to the same atomic mass (A): for instance, Xe, Te, and Sn have a stable isotope with $A = 124$. On the contrary, there are at most two (and generally only one) stable isotopes for odd atomic numbers and one isobar for odd atomic masses. This difference between nuclei with odd and/or even Z and/or A is strictly related to the greater stability of even nuclei; inside nuclei, neutrons behave like protons: they constitute two different states of charge of a same elementary particle, the nucleon, which is a fermion with two possible states of spin. There is a pairing effect between nucleons with different states of charge and spin: this is why nuclear species like ^{4}He, ^{12}C, ^{16}O, ^{20}Ne, ^{24}Mg ... are especially stable and abundant in nature. As it will be discussed later on in this chapter, the most stable and abundant nuclei are also those which have the largest number of isotopes.

Let us see now on the fragment of the chart of nuclides shown in Figure VI.2 what happens when nuclei experience the action of a neutron flux. When a nucleus absorbs a neutron it becomes a new isotope of the same element with an atomic mass ($A + 1$):

$$(Z, A) + n \rightarrow (Z, A + 1) + \gamma.$$

The nucleus $(Z, A + 1)$ can be stable or unstable against β decay. The β decay is the transformation of a neutron into a proton inside the nucleus by emission of an antineutrino and an electron (classically called β decay):

$$(Z, A + 1) \rightarrow (Z + 1, A + 1) + \bar{\nu} + e^-.$$

If the nucleus is stable it can itself absorb another neutron and become a heavier isotope of the same element. On the other hand, if the nucleus is unstable, the forthcoming nucleosynthetic path depends on the intensity of the neutron flux: *if the neutron flux is weak* the time elapsed between two neutron absorptions by a given unstable nucleus (which is $\sim 1/\Phi_n \sigma_n$ where Φ_n is the neutron flux and σ_n the neutron absorption cross section) is larger than the lifetime of the nucleus against β decay. This β decay lifetime generally ranges from a fraction of a second up to a few years. In this case (weak neutron flux), the nucleus has enough time to β decay before capturing another neutron. The nuclei formed by this process lie along the continuous path of stable nuclei in the chart of nuclides; they are called *s* elements according to their formation by *slow* neutron capture. If the mass M of the nuclei is added as a third dimension to the number Z of protons and to the number N of neutrons, (Figure VI.3) for a given isobar the stablest nuclei also have smallest mass. On this three-dimensional picture one can draw the 'valley of stability' where the stable nuclei can be found. The *s* process nuclei are obviously at the bottom of the 'valley of stability'.

Let us now assume that the *neutron flux is very intense*. In this case, the time scale against the neutron absorption is shorter than the β decay lifetime of the unstable nuclei: the nucleus $(Z, A + 1)$ captures another neutron before the β decaying which leads to the $(Z, A + 2)$ isotope:

$$(Z, A + 1) + n \rightarrow (Z, A + 2) + \gamma.$$

In this case, heavier and heavier unstable isotopes can be synthetized until the chain of neutron absorptions leads to such an unstable nucleus that the (γ, n) reaction occurs before the (n, γ) one. This rapid neutron process synthetizes *r* process elements which are located on the neutron rich side of the valley of stability. The path followed by the *r* process on the chart of the nuclides is represented in Figure VI.4.

As we shall see in the next section, the rate of neutron absorption $\langle \sigma_n v_n \rangle$ per

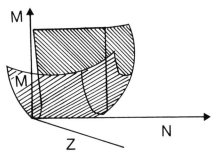

Fig. VI.3. Surface of the nucleus masses with respect to the number of protons Z and neutrons N. The stable nuclei lie in the so-called stability valley while the unstable nuclei which have larger masses are on the proton rich Z and neutron rich N instability hills.

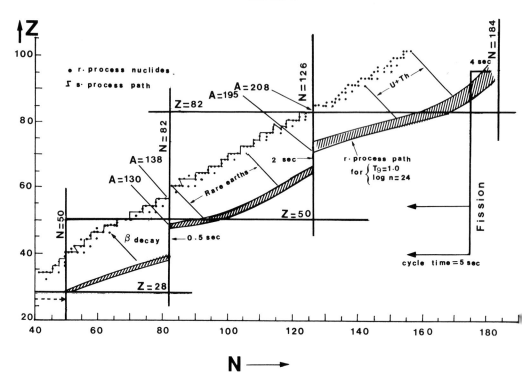

Fig. VI.4.   Neutron capture paths for the s process and the r process. The r process follows a path along the line of beta stability. The stable r process nuclei are shown here as small circles and have neutron rich progenitors, the waiting points, which are inside the shaded area (the r process path shown here has been calculated for $T = 10^9$ K and $n_n = 10^{24}$ neutrons cm^{-3}). Each magic number constitutes a discontinuity in the waiting point area which means that each stable 'magic' r process nucleus has more progenitors than its neighbors.

nucleus of heavy element is roughly constant. From the relations developed in Chapter IV, the lifetime of a heavy element nucleus against neutron absorption is given by:

$$\tau_n = \frac{1}{n_n \langle \sigma v \rangle}.$$

A good average estimate for σ is $\sim 10^{-25}$ cm^2 (i.e., 0.1 barn) while the thermal velocity is 3×10^8 cm s^{-1} which means that

$$\tau_n = \frac{10^9 \, \text{yr}}{n_n}.$$

The s process corresponds to neutron densities of $\sim 10^5$ neutrons cm^{-3} so that $\tau_n \sim 10^4$ yr is larger than any β decay lifetime, while for the r process the neutron density n_n should be $\sim 10^{22}$ neutrons cm^{-3} so that $\tau_n \sim 10^{-6}$ s shorter than the characteristic β decay lifetimes.

Three types of nuclei synthetized by neutron processes can be defined:

(1) The majority of them can be synthetized both by s or r process. They are called s, r nuclei.

(2) The pure neutron rich elements are stable but separated from the s path by at least one β unstable nucleus. (In Figure VI.2 this is the case for ^{116}Cd, ^{122}Sn, ^{124}Sn, ^{128}Te, ^{130}Te ... .) These are called r nuclei. It is important here to recall that the s process stops with ^{209}Pb. All the stable nuclei heavier than ^{209}Pb are also pure r process nuclei.

(3) The pure s nuclei are shielded from the r process by a neutron rich (r process) isobar: these pure s nuclei which cannot be reached by the r process are always even nuclei belonging to an isobaric family with one r process member. As we shall see in the sequel pure r process and pure s process nuclei are important to set up the astrophysical constraints on these two types of nucleosynthesis induced by neutron capture. Before going further, we can note that every peak corresponding to heavy elements in the SAD abundance curve (see Figure VI.1) is actually split into two humps. The lightest humps are due to the r process while the heaviest ones are due to the s process.

The stable isotopes which are on the proton rich side of the valley of stability cannot be synthetized by either process. They are called p process or by-passed elements. Their origin will be discussed in Section 5.

VI.2. Neutron Capture Reactions

The neutron capture reactions can be written:

$$(A, Z) + n \rightarrow (A + 1, Z) + \gamma.$$

Since there is no Coulomb barrier between the incoming neutron and the charged nucleus, these reactions occur much more easily than fusion reactions between charged particles. As a consequence, the neutron capture cross sections are much larger than the cross sections of fusion reactions. They are generally of the order of 100 to 1000 mb while the fusion cross sections are ~ 1 mb (10^{-3}). The ^{155}Gd(n, γ) ^{156}Gd reaction has a thermal cross section of 80 000 b while the geometrical cross section is of only 2 b: the heavy nuclei are particularly strong neutron absorbers. Contrary to the fusion reactions, the (n, γ) cross sections decrease with energy: $\sigma \propto (1/\sqrt{E})$ at very low (thermal) energies corresponding to $\sim$ a few eV, and $\sigma \propto (1\sqrt{E})$ at higher energies (in the keV region). In most astrophysical situations the $\langle \sigma v \rangle$ averaged product can be considered as roughly independent of the energy.

Figure VI.5 displays the (n, γ) cross section for a given energy against the atomic number of the heavy elements. The dips of the curve correspond to nuclei with proton or neutron 'magic numbers' which are especially stable. The concept of magic numbers used for nuclei in nuclear physics is similar to the concept of noble gas in atomic physics. In Chapter IV, it is seen that nucleons (protons and neutrons) could be considered as piled up in different shells within the nuclei similarly to the electrons in the atoms: the photons ejected after the neutron absorption are due to the arrangement of the nucleus among the different nucleon shells. The latter are filled up by nucleons in the same way as the electronic shells by electrons. When

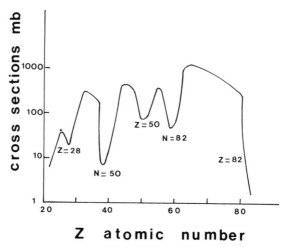

Z   atomic   number

Fig. VI.5.   Neutron absorption cross sections (n, γ), the hollow features of the curve are due to the fact that magic number nuclei are more stable than others (see text) and therefore, have smaller absorption cross sections than their neighbors.

the outer electronic shell of an atom is full, this atom has the configuration of a noble gas i.e. it is particularly stable against chemical reactions. In the same way, when the outer nuclear shells are filled up (which corresponds to the so-called magic numbers of protons and/or neutrons), the nucleus is especially stable against any nuclear reactions. The minima in the cross sections for magic number elements seen in Figure VI.5 correspond to the maxima or humps in the abundance curve displayed on Figure VI.1.

VI.3. The s Process

VI.3.1. THE MAIN NEUTRON SOURCES FOR THE s PROCESS

The s process is assumed to occur during the Red Giant phase of the stellar evolution i.e. during a stable phase of this evolution. Many nuclear reactions can be invoked to produce neutrons but the most important are those which take place during the He burning (Chapter IV). During the H burning which occurs when the star is on the Main Sequence, the only reaction able to release some neutrons is $D + D \rightarrow {}^3He + n$; but it competes with $D + p \rightarrow {}^3He + \gamma$ and $D + D \rightarrow {}^3He + p$. Moreover, the neutrons produced by this reaction are rapidly absorbed by ${}^3He$ and ${}^{14}N$, nuclei which are largely formed respectively by the H burning chain PPI (see Chapter IV) and the CNO cycle. They are important poisons for neutrons because the ${}^3He(n, p){}^3H$ and ${}^{14}N(n, p){}^{14}C$ have large cross sections.

   Neutrons are much more easily released during the Red Giant phase, either during flashes which allow an incomplete mixing of the H and He zones or in the He burning zone itself.

   (a) The H burning zone can be enriched into ${}^{13}C$ as well as into ${}^{14}N$ especially if there is an incomplete mixing between this zone and the He burning zone. If the mixing of He and H burning material is such that the abundance in ^{12}C in the mixed matter is larger than the H one, the CNO cycle is incomplete and reduces to the reactions $^{12}C(p, \gamma)^{13}N(^+\beta)^{13}C$. ^{13}C itself can react with He and provide an important source of neutrons through $^{13}C(\alpha, n)^{16}O$ which has a rather large cross section. Furthermore, in this case, ^{14}N is not significantly synthetized and therefore is not a poison for the neutrons.

(b) In the He burning zone the ^{14}N which was present before the nuclear processing can itself react with He and induce the following chain of reactions:
$^{14}N(\alpha, \gamma)^{18}F(\beta^+)^{18}O(\alpha, \gamma)^{22}Ne(\alpha, n)^{25}Mg$; $\quad$ $^{18}O(\alpha, n)^{21}Ne$ ... $\quad$ Between $\quad$ the $^{13}C(\alpha, n)^{16}O$ reactions and those induced from the He burning of ^{14}N; it is not known yet which is the most important neutron source. These reactions are able to produce up to 10^5 neutrons per Fe atom at temperatures of $\sim 10^8$ K.

Finally, let us mention the reactions which occur at higher temperatures during the C and O burning: $^{12}C + {}^{12}C \rightarrow {}^{23}Mg + n$ which releases about 10 neutrons per Fe atom at 10^9 K while the $^{16}O + {}^{16}O \rightarrow {}^{31}S + n$ releases about 100 neutrons per Fe atom at some higher temperatures.

VI.3.2. THE s PROCESS NUCLEOSYNTHESIS

As it has been seen in the two previous chapters, the effect of any nucleosynthetic process acting on a nuclear species A can be estimated by solving the differential equation describing the abundance variation of the element A with time:

$$\frac{dN_A}{dt} = - N_A N_B \langle \sigma v \rangle_{AB} + N_C N_D \langle \sigma v \rangle_{CD}.$$

In general, when different nuclear reactions can synthetize and/or destroy the same element, one has to solve a system of differential equations for which the computations can be complicated, see Chapter V.

For the s process, the equation $N_A(t)$ which describes the time variation of the abundance of a heavy element of atomic mass A, can be simply written as:

$$\frac{dN_A(t)}{dt} = - \langle \sigma v \rangle_A n_n(t) N_A(t) + \langle \sigma v \rangle_{A-1} n_n(t) N_{A-1}(t). \tag{VI.1}$$

According to the previous section, the absorption cross section and the neutron velocity do not vary much during this process. If ϕ is the neutron flux integrated on the duration of the process we have:

$$d\phi = n_n(t) \, v(t) \, dt$$

and Equation (VI.1) then becomes:

$$\frac{dN_A}{d\phi} = - \sigma_A N_A + \sigma_{A-1} N_{A-1}, \tag{VI.2}$$

where N_A and N_{A-1} are the abundances of the elements A and $A - 1$ and σ_A and σ_{A-1} the cross sections of $A(n, \gamma) (A + 1)$ and $(A - 1) (n, \gamma) A$.

Equation (VI.2) is valid for any element such as $57 < A < 206$.

It can be assumed that the seed for the s process is ^{56}Fe. Moreover, the s process stops occurring at ^{209}Pb$(n, \alpha)^{206}$Pb; the first and the last equation of the system are then respectively:

$$\frac{dN_{56}}{dt} = -N_{56}\,\sigma_{56}$$

and

$$\frac{dN_{206}}{dt} = -N_{206}\sigma_{206} + N_{205}\sigma_{205} + N_{209}\sigma_{209}.$$

If $\sigma_A N_A < \sigma_{A-1}N_{A-1}$, dN_A/dt is positive and N_A increases while if $\sigma_A N_A < \sigma_{A-1}N_{A-1}$, N_A decreases. The abundances should then reach an equilibrium such that:

$$\sigma_A N_A = \sigma_{A-1}N_{A-1}$$

provided that the time scale for the s process to take place is short enough. This very simple result is in fact one of the best proofs in favor of the s process. It is illustrated by Figure VI.6, where the product σN for the s process elements is plotted as a function of the atomic mass: the humps in the abundance curve (Figure VI.1) correspond to the dips in the cross section curve (Figure VI.5). These humps and dips corresponding to the magic numbers vanish in Figure VI.6 where the curve σN is plotted. It is interesting to compare the smoothness of the σN curve corresponding to the s process elements to the erratic picture displayed by the σN product for the r process elements (Figure VI.7). This constitutes a proof that these elements are formed by the s process while the r elements are formed by another process. The abundance curve for the s process appears to be smoothly decreasing with the atomic mass A. The reason for this decrease is that the equilibrium $\sigma_A N_A = \sigma_{A-1}N_{A-1}$ has not yet been reached (otherwise the σN line would be parallel to the axis A). The lighter the elements the closer they are from their equilibrium abundance. The best model to account for the observed abundance distribution is to assume that the seed for the s process (i.e. the Fe nuclei) has experienced a series of different neutronic irradiations: a few long neutron irradiations for a small number of Fe nuclei and many short irradiations for a large fraction of them. These short irradiations are responsible for the formation of the light s process nuclei while the long ones are responsible for the formation of the heaviest.

As it is said throughout this monograph, the surface of a given star generally does not reflect the abundances of the elements synthetized in the internal regions of this star. However, the characteristic spectroscopic lines of Tc, a radioactive element with a lifetime $\sim 10^5$ yr have been observed in the spectrum of several cool stars. This element should come from the star itself the age of which is much longer than the Tc lifetime. Finally, as mentioned in Chapter V, one of the most exciting and recent discoveries is the detection of two variable stars FG Sagittae and CI Cygni.

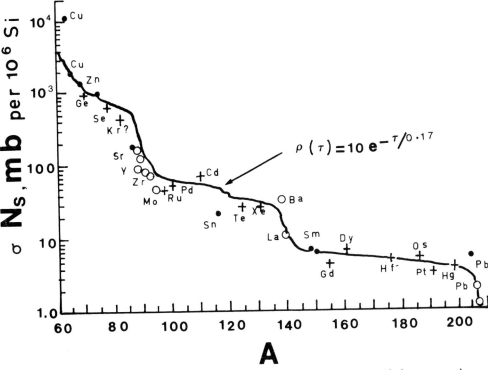

Fig. VI.6.   The solar σN_s curve (product of the neutron absorption cross section with the s process abundances) against the atomic mass A. The cross sections are those measured at 30 keV and expressed in mb, while the nuclide abundances are given assuming that the abundance of ^{28}Si is 10^6. One notes that the bumps and hollows of Figure VI.1 and VI.5 have disappeared. The continuous aspect of this curve is the best proof of the occurrence of the s process nucleosynthesis inside stars. The solid line is a theoretical calculation due to Clayton *et al.* (1961).

FG Sagittae not only shows a significant decrease of its superficial temperature (~ 250 K yr^{-1}) but also an annual increase by a factor ~ 3 of the abundances of the s process elements such as Zr, Y, Ba observed during a few years around 1970 to 1973. CI Cygni seems to show an important increase of rare earth elements, also formed by the s process. These two stars play a very important role in nuclear astrophysics because they really show the actual manifestation of a nucleosynthetic process going on in a star.

VI.4. The r Process

The r process nucleosynthesis involves heavy elements with a neutron rich nucleus. It is not as well established as the s process nucleosynthesis. The neutron absorption leads to a r process nucleosynthesis only if the neutron captures occur on time scales shorter than the β decays. The corresponding neutron fluxes should be then as

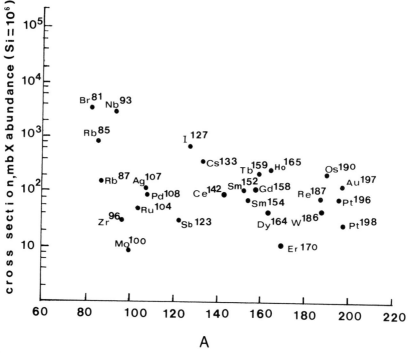

A

Fig. VI.7. The product of the neutron absorption cross section times the nuclide abundance for the r process nuclei. Contrary to the s process (Figure VI.6), one notices the large irregularities in the product σN. This means that the smoothness of the σN curve of the s process is not accidental!

high as $\Phi_n \sim 10^{22}$ to 10^{26} neutrons cm^{-2}. Up to now, the supernovae are the only astrophysical objects which seem able to release large neutron fluxes.

In supernovae some internal zones can reach very high temperatures and densities such as $T > 10^9$ K, $\rho > 10^6$ g cm^{-3}. In these zones, protons can be transformed into neutrons by electron absorption (p + e$^-$ → n + γ). These neutrons could be subsequently accelerated by the shock waves triggered by the supernova explosion, and rapidly be absorbed by the heavy elements contained in its outer zones.

Although the r process is still poorly known there cannot be any doubt on its occurrence since elements such as Th, U or Pu cannot be synthetized by any other mechanism. Figure VI.8 shows the abundances of the r process elements determined by subtraction of the s process abundances from the SAD abundances (see Chapter II). In this figure, one can notice the three peaks corresponding to the magic number nuclei Se, Te and Pt* and between the Te and Pt peaks, the 'hill' of the rare earths (around Gd).

The r process nucleosynthetic path is very different from the s process one. It lies in a zone of the Z, N plane where the nuclei which are very neutron rich, are generally very unstable. Neutron capture only stops when the probability becomes larger

* The number of neutrons inside the corresponding nucleus is respectively 50, 82, 126.

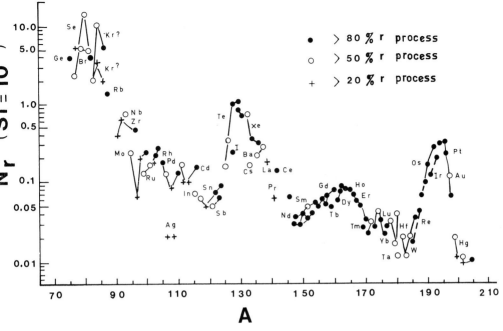

Fig. VI.8.   The solar system r process abundances. The r process abundances have been obtained by subtracting the s process contribution from the total abundance of the nucleus. The most important characteristics are the three main abundance peaks (Se, Te, Pt) and the broad 'hill' of the rare earths (Gd, Ho).

for the (γ, n) reaction than the (n, γ) one. This occurs when the binding energy of the last neutron in the $(Z, A + 1)$ nucleus is so weak that the (Z, A) nucleus cannot absorb this neutron. This point on the (Z, N) plane is called a waiting point. The nucleus undergoes no reaction until a β decay occurs which makes its atomic number increase by one unit from Z to $Z + 1$. Then the neutron absorption proceeds on the atomic number $(Z + 1)$ and so forth . . .

Under these conditions, the relative abundance of each sequence of nuclei Z is given by the following equation:

$$\frac{d}{dt}\left(n_Z(t)\right) = \lambda_{Z-1}n_{Z-1}(t) - \lambda_Z n_Z(t),$$

where λ_{Z-1} and λ_Z are respectively the rates of β decay of the waiting points corresponding to the charges $Z - 1$ and Z.

Between the different magic numbers, the curve of the waiting points is rather smooth (Figure VI.8). For $N = 50$, 82, and 126 (corresponding to the respective magic numbers) the curve of the waiting points becomes almost vertical. Consequently, the stable r nuclei corresponding to forementioned magic numbers have many more parents than the other r nuclei. This explains the peaks in the abundance curve.

The position of the waiting points of the r process depends on both the strength of the neutron flux and the temperature of the medium where the r process takes place. Larger neutron fluxes favor more neutron rich waiting points while if the temperature of the medium increases the waiting points are less neutron rich. The relative abundances of the r processes nuclei depend strongly on the mass of these nuclei (or on the binding energy of the last neutron inside these nuclei). Therefore, in order to understand the r process nucleosynthesis, one should know precisely the binding energies (i.e. the masses) of the very neutron rich nuclei located very far from the valley of stability and thus not observable in nature. Then, the estimate of the mass of these nuclei can only be made on theoretical grounds.

The tables of theoretical masses are based on representations of the still unknown nuclear forces binding the nucleons inside the nucleus. Theoretical nuclear physicists are trying to build up nuclear mass formulae which fit as well as possible the masses of the measurable nuclei. For a long time, the old but classical formula of Weiszäcker based on a liquid drop representation of the nucleus has been used. From this model, the binding energy $B(Z, A)$ is given by:

$$B(Z, A) = \left(\alpha - \frac{\gamma}{A^{1/3}} \right) A - \left(\beta - \frac{\eta}{A^{1/3}} \right) \frac{I^2 + 4(I)}{A} +$$

$$+ \delta + \left(\pm \frac{1}{A^{1/2}}, 0 \right) - 0.8 \frac{Z^2}{A^{1/3}} \left(1 - \frac{0.76}{Z^{2/3}} \right) \left(1 - \frac{2.00}{A^{2/3}} \right).$$

In this formula the neutron excess of the (A, Z) nucleus is: $I = N - Z$; α, β, γ, δ, and η are five parameters determined from the masses of observable nuclei. The term αA expresses the dependence of the mass of a nucleus with the total number of nucleons inside it. The term $\gamma A^{2/3}$ is a liquid drop term of superficial strain related to the non saturation of the nuclear forces exerted on the nucleons at the surface of the nucleus; the term in $Z^2/A^{1/3}$ expresses the effect of the Coulomb forces applied by the protons inside the nucleus. The term depending on I implies that a nucleus is more stable if the difference $N - Z$ is small, while the term depending on δ means that the even Z and even N nuclei are more stable than the odd A nuclei which are themselves more stable than the odd Z odd N nuclei.

Actually, it appears that this old formula is not satisfactory enough to be applied to the r process. Progress on this nucleosynthetic process will be made only if better mass formulae can be designed, which needs a better understanding of the nature of the nuclear force. A tremendous effort is devoted in many laboratories to try to solve this difficult question: for example, let us mention in France the groups of theoretical nuclear physicists in Orsay and Bruyères-le-Chatel. For the moment, one can only say that the r process is not yet fully understood.

VI.5. The p Process

The p process elements are stable but much less abundant than the s and r process ones (by a factor ranging from 0.001 to 0.01). In first approximation their abundance curve is parallel to the s and r element curve. In particular, the p process elements with magic number nuclei, are more abundant than their neighbors. However, the ratio of the abundances of the p over r and s elements for a given A decreases with A.

These elements cannot be synthetized by neutron absorption processes. All the mechanisms which are invoked to explain their formation assume that the elements are derived from the s and r elements (which can account for their magic number peaks).

These mechanisms are as follows:

VI.5.1. WEAK INTERACTION MECHANISM

At temperatures higher than 10^9 K photons which have an energy $E > 1$ MeV can be transformed into an $e^- e^+$ pair

$$\gamma \rightarrow e^+ + e^-, \qquad (Q \sim 1.022 \text{ MeV}).$$

The positrons which are released by such a transformation can be absorbed by nuclei and then transform neutrons into protons according to the following weak interactions:

$$e^+ + (A, Z) \rightarrow (A, Z + 1) + \bar{\nu}$$

$$\gamma + (A, Z) \rightarrow (A, Z + 1) + e^- + \bar{\nu}.$$

This photon absorption corresponds to an implicit positron absorption. In Figure VI.9, this mechanism corresponds to the transformation of the s (or r) nucleus (2) into the p nucleus.

This mechanism can occur in any zone where $T > 10^9$ K (for instance, inside the O burning zone). However it cannot account for the formation of all the p process elements. The rates of these weak interactions vary largely indeed from one element to its neighbors and cannot reproduce the smoothness of the p process abundance curve. Nevertheless they might be responsible for the formation of a p process element, ^{164}Er, which is particularly abundant and for which the corresponding rate of weak interaction is indeed especially high.

VI.5.2. SPALLATION REACTIONS

As developed in Chapter VII, spallation reactions are endothermic processes; they consist in a partial destruction of the complex target nuclei. These processes are generally induced by energetic incoming particles ($E > $ a few MeV). The p process elements could be formed by such spallation reactions according to the following scheme:

$$(A, Z) + \text{p} \quad \text{or} \quad \alpha \rightarrow (A', Z') + \dots$$

where $A' < A$ and $Z' < Z$.

These mechanisms can take place when energetic protons or α particles are present, at the passage of a shock wave throughout the surface of a supernova or in the case of interaction of the Galactic Cosmic Rays with the interstellar medium for example.

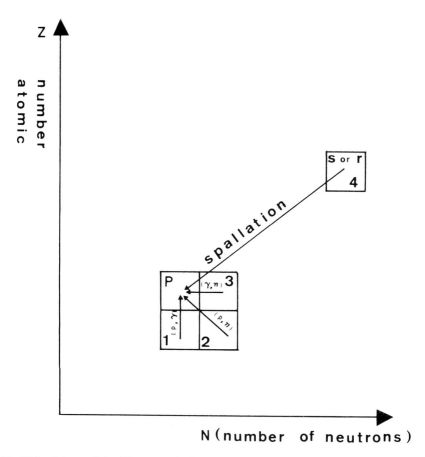

Fig. VI.9. Scheme of the different mechanisms able to produce a *p* process nucleus from *s* or *r* process nuclei. The *s* or *r* nucleus (1) is transformed into *p* by thermonuclear (p, γ) reaction; (2) is transformed into *p* either by (p, n) reaction or by weak interaction (positron absorption or photon absorption followed by an electron emission); (3) is transformed into *p* by (γ, n) reaction; Finally, (4) can be changed into *p* by spallation reactions (discussed in Chapter VII).

These spallation mechanisms can only explain the formation of very underabundant *p* process elements like ^{138}La or ^{180}Ta which are *Z*-odd *N*-odd nuclei. Even nuclei (*A* even) can have either even *Z* and even *N* or odd *Z* and odd *N*. The *Z*-odd *N*-odd nuclei are much less stable than the *Z*-even *N*-even nuclei and therefore have much lower abundances; these *Z*-odd *N*-odd nuclei are much less abundant than the other *p* process *Z*-even *N*-even elements. This spallation mechanism might also be responsible for the formation of the heaviest *p* process elements such as ^{184}W, ^{190}Pt, ^{196}Hg which might come from the spallation of Pb, particularly abundant in this atomic mass range.

VI.5.3. THERMONUCLEAR REACTIONS

Two types of thermonuclear reactions can transform an s (or r) element into a p process one. They are:

(a) (p, γ) reactions transforming the nucleus (1) into p (Figure VI.9):

$$(A, Z) + p \rightarrow (A + 1, Z + 1) + \gamma.$$

(b) (γ, n) reactions transforming the nucleus (3) into p (Figure VI.9):

$$\gamma + (A, Z) \rightarrow (A - 1, Z) + n.$$

These reactions can occur at temperatures $T \sim 10^9$ K. The (p, γ) reactions may affect the low mass range ($A < 150$) for which the Coulomb barrier which is proportional to Z is relatively weak. The (γ, n) reactions transform heavy s or r nuclei ($A > 150$) into heavy p nuclei.

The mechanisms based on these thermonuclear reactions are indeed the most promising ones to account for the synthesis of such elements. Up to now, explosive nucleosynthetic processes occurring either in H or O rich zones have been considered in this respect. One of us (J.A.) and J. W. Truran have shown that explosive nucleosynthesis (see Chapter V) occurring in H rich zones (material with solar abundance) at temperatures $T \sim 2 \times 10^9$ K and densities of $\sim 10^4$ g cm^{-3} could explain the formation of about 2/3 of the observed abundances of the p elements. The remaining exceptions could be due to the great uncertainty with which the rates of thermonuclear reactions and β decays are known. However, one should be aware that such a model is far too simple for astrophysics. Except for extreme models of novae where the maximum temperature in the explosive zone can be as high as 2×10^9 K there are no known astrophysical conditions where H rich material can reach such high temperatures and densities. This is why models based on nucleosynthesis occurring in O rich zones have also been developed. But as well as the H zone model, this mechanism is not free from difficulties especially when one considers the number of free parameters needed to account for the observed abundances.

To conclude on the p process, it is fair to say that like the r process, this type of nucleosynthesis is not yet fully understood.

To end this chapter devoted to the nucleosynthetic processes able to explain the formation of the heavy elements, the same remark can be made as for the other nucleosynthetic processes: the general scheme is undoubtedly a correct description of the phenomenon but the final details of the three processes are far from being correctly accounted for.

References

Quoted in the text:

Cameron, A. G. W.: 1973, *Space Sci. Rev.* **15**, 121.
Clayton D. D., Fowler, W. A., Hull, T. E., and Zimmerman, B. A.: 1961, *Ann. Phys.* **12**, 331.

Further readings:

Arnett, W.D., Hansen, C.J., Truran, J.W., and Cameron, A.G.W. (eds.): 1968, *Nucleosynthesis*, Gordon and Breach Science Pubs., Inc., New York.
Audouze, J. and Truran, J. W.: 1975, *Astrophys. J.* **202**, 204.
Hillebrandt, W.: 1978, *Space Sci. Rev.* **21**, 639.
Iben, Jr. I.: 1975, *Astrophys. J.* **196**, 525; **196**, 549.
Schramm, D.N. and Arnett, W.D.: 1973, *Explosive Nucleosynthesis*, University of Texas Press, Austin.
Seeger, P.A., Fowler, W.A., and Clayton, D.D.: 1965, *Astrophys. J. Suppl.* **11**, 121.
Truran, J.W. and Iben, Jr. I.: 1977, *Astrophys. J.* **216**, 797.
Ulrich, R.K.: 1973, in D.N. Schramm and W.D. Arnett (eds.), *Explosive Nucleosynthesis*, University of Texas Press, Austin.

CHAPTER VII

NUCLEOSYNTHESIS OF THE LIGHT ELEMENTS

As seen in the previous chapters, the light elements D, Li, Be and B are not produced by stellar nucleosynthesis. These elements are destroyed by (p, γ) or (p, α) reactions occurring at temperatures higher than $\sim 10^6$ K. For instance, in the H burning cycle the $D(p, \gamma)^3$He reaction is so rapid compared to the other chains of the cycle that the D/H equilibrium ratio is $\sim 10^{-18}$ i.e., 10^{-13} times the observed abundances of D. In the normal course of the nucleosynthetic events occurring inside the stars, one goes directly from the production of ^{4}He to that of ^{12}C and ^{16}O by-passing the elements Li, Be and B.

Contrary to the other light elements, He is formed in Main Sequence stars. Nevertheless, its observed abundance is too large to be accounted for by stellar nucleosynthesis only. One generally considers that only 20 to 30% of the observed He (He/H ~ 0.10) is produced by the stars. The abundance of ^{3}He has only been measured in the solar wind and in some peculiar stars (e.g., 3 Cen A). Although the present interstellar ^{3}He abundance is not accurately determined, a value ^{3}He/H ~ 1 to 2×10^{-5} is generally adopted.

It is then necessary to invoke other nucleosynthetic processes to explain the presence of these nuclear species. Many different mechanisms have been studied. It has been found that the elements Li, Be and B are mainly formed by endothermic reactions (the spallation reactions) which occur at much higher energies ($E >$ a few MeV) than the fusion reaction (Section VII.2). The spallation reactions result from the interaction between the Galactic Cosmic Rays and the interstellar medium (Section VII.3). Some of these elements (basically ^{7}Li) can also be formed in stellar events (Section VII.4) while D, ^{3}He and ^{4}He are believed to have been formed during the Big Bang (Section VII.5). Let us first summarize the results concerning the observed abundances of the light elements (Section VII.1).

VII.1. The Abundance of the Light Elements

Table VII.1 gives an updated excerpt of the present observations of the light element abundances. The most striking features are the following:

(1) At first glance, the D abundance seems to be the same in the Solar System as in the local interstellar medium (D/H ~ 1–2×10^{-5})*. However, recent observations made in the UV in front of hot O and B stars with the Copernicus (OAO-3) satellite, seem to show some dispersion of the D/H ratio determined in different lines of sight. This dispersion can be explained by physical effects. Furthermore, recent

---

* Here one should recall that D which is burnt inside the stars has not been found in stellar surfaces.

TABLE VII.1

Compendium of the abundances of the light elements $A \leq 11$

	Interstellar matter	Stars	Solar surface	Solar system	Meteorites	Adopted galactic
D	$10^{-6} \to 2 \times 10^{-5}$	$< 10^{-6}$	$< 2.5 \times 10^{-7}$	$(2 \pm 0.4) \times 10^{-5}$	1.3 to 2×10^{-4}	1 to 2×10^{-5}
^{3}He	$< 4 \times 10^{-5}$			$(2 \pm 1) \times 10^{-5}$		1 to 2×10^{-5}
^{4}He	0.06 to 0.10	~ 0.10				0.1
Li	3 to 6×10^{-10}	$10^{-7}\dagger - 10^{-9}\ddagger$	10^{-11}		1.5×10^{-9}	$10^{-9} \pm 0.3$
Be	$< 7 \times 10^{-11}$	1 to 2×10^{-11}	10^{-11}		2×10^{-11}	10^{-11}
B	$< 2 \times 10^{-9}$	1.5×10^{-10}	1.5×10^{-10}		$3 \times 10^{-10} \to 1.5 \times 10^{-9}$	$3 \times 10^{-10 \pm 0.3}$

† Red Giants   ‡ Main Sequence

The Li and Be isotopic ratios in the Solar System are respectively

$$\frac{^7\text{Li}}{^6\text{Li}} = 12.5 \qquad \frac{^{11}\text{B}}{^{10}\text{B}} = 4.0.$$

determinations of the DCN/HCN ratio in various molecular clouds (Orion, W51 and especially the molecular clouds Sgr A and Sgr B2 in the central regions of the Galaxy) seem to indicate that D has a lower abundance in the Galactic Center: this effect will be explained in Chapter IX in terms of galactic evolution. Here, we adopt a present D/H ratio of 1 to 2×10^{-5}.

(2) In the case of Li, it appears that all the youngest stars such as the T Tauri stars have roughly the same abundance (Li/H $\sim 10^{-9}$). This abundance is close to the abundance of the solar system (Li/H $\sim 1.2 \times 10^{-9}$) as well as to the one found in the local interstellar medium (Li/H ~ 3 to 6×10^{-10}). The Li abundance decreases at the surface of older stars: for instance, at the surface of the Sun Li/H $\sim 10^{-11}$. Li is less abundant in stars with a cooler surface and presumably a deeper external convective zone (Figure VII.1). This is probably due to the thermonuclear destruction of ^{6}Li and ^{7}Li induced by protons in stellar interiors. This correlation between the Li abundance and the age of the stars is only true for Main Sequence stars: a few Red Giant stars happen to show Li/H $\sim 10^{-7}$. Finally, the isotopic ratio ^{7}Li/^{6}Li has been observed to be larger than 10. (In the Solar System ^{7}Li/^{6}Li ~ 12.5.)

(3) Be has been observed with the same abundance (Be/H $\sim 2 \times 10^{-11}$) in most of the Main Sequence stars in which it has been searched for as in the Solar System. Be is 10 times underabundant only in some F stars with $T_{eff} > 6600$ K.

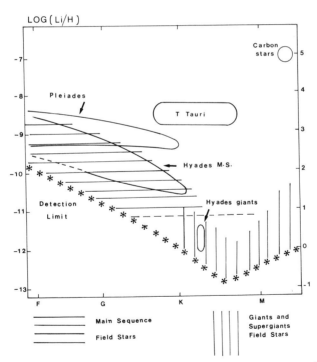

Fig. VII.1. Li abundance with respect to the spectral type of stars. One notices an obvious decrease of Li/H with spectral type going from F to K (decrease of the superficial temperature). T. Tauri stars are pre Main Sequence stars while carbon stars are very evolved stars. They do not follow this trend which applies mainly to Main Sequence stars.

(4) The abundance of B has been considered as a puzzle for the last few years: on one hand the B/H ratio 10^{-9} only uses carbonaceous chondrites. On the other hand B/H is found to be only a few 10^{-10} at the solar surface, in the nearby interstellar medium and at the surface of a few normal stars. Now, one generally chooses the lower value B/H $\sim$ a few $\times 10^{-10}$ as the normal abundance, assuming that chemical fractionation processes are responsible for the high value in the case of carbonaceous chondrites. (Similar discrepancies have also been found between the solar and the meteoritical value for other volatile elements, like Hg.)

VII.2. The Spallation Reactions

Spallation reactions are high energy and endothermic nuclear reactions which consist in a partial destruction of a heavy nucleus experiencing a shock at high energy ($E >$ a few MeV at least) with light projectiles such as protons or α particles. In what follows, we call 'projectile' the light particle which is assumed to bombard the heavy target at rest (the situation can be reversed and describes as well the collision between rapid heavy nuclei and light nuclei at rest).

Figure VII.2 shows an example of some experimental cross sections of the spallation of ^{12}C induced by protons: it indicates the characteristic energy range of the spallation reactions which should occur at least at a few tens MeV (i.e. at energies 10^3 times larger than thermonuclear energies).

The characteristic features of these reactions can be seen on Figure VII.2. These reactions are endothermic (see Chapter IV and Appendix C).

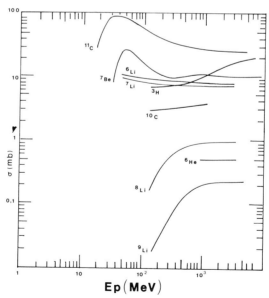

Fig. VII.2.   Experimental cross section for the production of various isotopes from ^{12}C + p.

The shape of the excitation function $\sigma(E)$ strongly depends on the value of the threshold E_p: if E_T is $\lesssim 20$ MeV, $\sigma(E)$ rises up to a rather steep maximum E_p ($E_p = E_T + 20$ MeV) and then decreases rather sharply down to an asymptotic value reached at $E \sim 100$ to 150 MeV. When the threshold is higher, the excitation function becomes flatter and the relative height of the maximum decreases. In fact when the reaction is highly endoenergetic, this maximum disappears and the cross section then increases smoothly before reaching its asymptotic value at about the same energy ($E \sim 100$ to 150 MeV).

This shape of $\sigma(E)$ has been understood for about 30 yr in the frame of the rather simple and now classical Serber model. The Serber model works in two steps: a) while the projectile crosses the target nucleus (in 10^{-21} to 10^{-22} s) the incoming particle collides with a few nucleons inside the target. These nucleons in turn are able to generate more collisions then creating a *cascade*. Some of the nucleons may have enough energy to leave the target nucleus with the projectile; and b) after this first rapid step a rather 'hot' and excited nucleus remains which dissipates its energy more slowly (in $\sim 10^{-16}$ s) by emitting a few other nucleons during this longer phase called the 'break-up'. Figure VII.3 gives a sketch of this simple but meaningful picture.

One can then understand, at least qualitatively, the trend of the excitation functions $\sigma(E)$ described above: the increase of $\sigma(E)$ at a low energy is due to the relative increase of the volume in the phase space when $E > E_T$. The maximum cross section is reached when the spalled nucleus can begin to release its energy through other channels, each of them representing a different spallation reaction. When E increases further, the other channels (or possible spallation reactions) compete and take a larger share of the available volume in the phase space. The asymptotic value of the spallation cross section is reached when there are more channels to be open.

For astrophysical purposes it has been found useful to develop semi-phenomenological or semi-empirical formulae to provide reasonable estimates of the spallation cross sections when no experimental measurements are available. In the case of spallation reactions induced by protons these phenomenological formulae

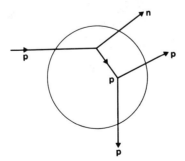

Fig. VII.3. Sketch of the first step of the spallation reactions ('cascade'). The incoming particle penetrates inside the nucleus and collides with a few nucleons before going out (after 10^{-22} to 10^{-21} s) together with the bombarded nucleons. The residual nucleus is very excited and ready to undergo the second slower step ($\sim 10^{-16}$ s) of nucleon evaporation.

have the following schematic form (as shown on Figure VII.4):

$$\sigma = \sigma_0 \exp\left(-P\Delta A - RT_p^2\right).$$

In this expression, the parameters P and R are decreasing functions of the energy, σ_0 is a constant parameter which depends only on the nature of the target and which has the dimension of a cross section. The term ΔA represents the mass difference between the target and the product: for instance in the case of $^{12}C + p \rightarrow {}^7Li + 2p + {}^4He$, $\Delta A = 5$. Finally T_p is a symmetry factor

$$T_p = \frac{|N_p - Z_p|}{2},$$

where N_p and Z_p are respectively the number of neutrons and protons of the product nucleus (in the previous example $T(^7Li) = \frac{1}{2}$).

The term $e^{-P\Delta A}$ is mainly related to the cascade: it is easier for an incoming particle to eject a few nucleons from the target than many. The term $\exp(-RT_p^2)$ is mainly related to the second phase or 'break up': it is more probable to end up with a more stable final nucleus (which has about the same number of protons and neutrons).

These rules and formulae are very useful in the prediction of cross sections for the spallation reactions leading to the production of rare light elements like Li, Be and B out of more abundant heavier nuclei (mainly C, N, and O).

VII.3. Production of Li, Be, B by the Galactic Cosmic Rays

The origin of the light elements in terms of spallation reactions has been considered for about 20 yr. It was first believed that these elements were produced in stellar flares occurring mainly when the stars are young, i.e. during the active T Tauri phase. This hypothesis has been abandoned because these spallation processes would demand an amount of energy much larger than what can be effectively released by these young stars in which there is still no fusion reactions. This conclusion

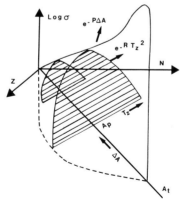

Fig. VII.4. Distribution of the experimental average spallation cross sections in the Z–N plane as a function of $\Delta A(A_T - A_p)$ and $Tp\ [(N_p - Z_p)/2]$ for a symmetrical target of mass $A_T\ (T_T = 0)$.

prompted Reeves *et al.* in 1970, following a suggestion of Peters, to propose the so-called 'galactogenic' hypothesis: The formation of the light elements is due to the interaction between the Galactic Cosmic Rays (GCR) and the interstellar medium. The GCR designate the high energy particles (10^6 to 10^{20} eV) coming from outside the Solar System. GCR particles have to be studied in balloon and satellite experiments because of their strong interaction with the terrestrial atmosphere which modifies them before they reach the ground. This interaction is used to detect the very high energy GCR particles ($E > 10^{15}$ eV). At these energies, the GCR particles induce in the atmosphere important showers of secondary particles (electrons, pions, muons etc...) from which the energy of the primary GCR particles can be deduced. It is in the GCR that the first elementary particles (positrons in 1932, pions in 1947) were discovered.

The characteristics of the GCR are the following:

(1) The GCR flux is isotropic and time independent.

(2) The fluxes decrease with energy according to a spectrum in power law of the energy:

$$\phi\,(E > \text{a few GeV/nucleons}) \propto E^{-2.7},$$

where E is the energy of the GCR particles (per nucleon).

(3) At low energy ($E < 1$ GeV/nucleon) the GCR flux is significantly affected by the 'solar modulation' (Figure VII.5): in the solar cavity (which roughly corresponds to the volume occupied by the Solar System), the solar wind and the solar magnetic field interact with the low energy cosmic rays; they prevent a large part of the low energy GCR particles from entering this cavity; the large fluxes of particles observed at $E \sim 1$ to 10 MeV are of solar origin and come from the Sun itself (they are called Solar Cosmic Rays). The solar origin of the modulation of the low energy flux can be proved by the fact that the intensity of this effect is strictly related to the 11 yr cycle of the solar activity. One does not know yet the quantitative importance of the solar modulation. Nevertheless, experiments embarked on presently launched satellites, such as Voyager, will provide measurements of the GCR fluxes at distances as large as Jupiter's and Saturn's orbits.

(4) The GCR composition is rather different from the cosmic (or universal) abundance (Figure VII.6).

In the GCR:

(i) The proportion of heavy elements like C, O and Fe compared to H or He is 10 to 50 times larger than in the Solar System.

(ii) The proportion of odd Z (Z = atomic number) elements compared to even Z nuclei for example (V/Fe, K/Ca ...) is larger in the GCR than in the cosmic abundances.

(iii) The ratio of light elements Li, Be, B against the so-called M elements C, N, O, is Li Be B ~ 0.23 in the GCR i.e. more than 10^4 times the value of this ratio in the cosmic abundances.

These two last features of the GCR composition are the signature of the interaction between the cosmic rays and the interstellar medium: between their source, which is generally assumed to be objects like supernovae able to release high energy particles, and the Solar System, the high energy cosmic rays travel inside the inter-

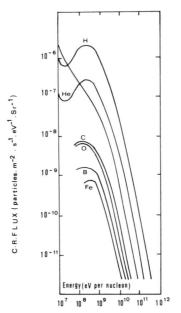

Fig. VII.5.   Dependence on the energy of the cosmic ray flux (given in number of particles per m² per second per steradian and within an energy range of 1 MeV). The maximum of the flux is located at $E = 1$ GeV/nucleon. Below this energy, the cosmic ray flux is affected by the solar modulation; this modulation effect is caused by the interplanetary magnetic field and the solar wind and largely prevents the low energy cosmic rays to penetrate the solar cavity. This effect varies with time following the 11 yr cycle of the Sun. Above 1 GeV/nucleon, the cosmic ray spectrum varies like $E^{-2.6}$ (E being the energy per nucleon) and does not vary with time. In contrast, below 20 MeV, the protons and the α particles come from the solar flares and vary largely with time.

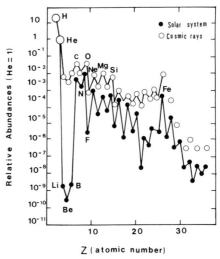

Fig. VII.6.   A comparison of the abundance of elements in cosmic rays and in the Solar System. The solar H abundance is normalized to 10^{12}, and the cosmic ray abundances are normalized to the solar values for magnesium. The abundances of Li are the meteoritic ones, because Li has burned up in the Sun.

stellar medium. They do not follow a straight trajectory: they are randomized and isotropized by the interstellar magnetic fields (of a few 10^{-6} to 10^{-5} G intensity) which bend the GCR trajectories. On this path, they hit interstellar atoms with which they interact by spallation processes creating odd nuclei and light elements. These elements are called secondary nuclei in contrast to the primary nuclei (C, O, Fe ...) which are produced in the GCR sources.

(iv) One should finally note that the ratio of the secondary elements against the primary elements decreases at high energies ($E > 10$ GeV/nucleon). This means that the high energy GCR particles ($E > 10$ GeV/nucleon) encounter less interstellar matter than GCR particles of lower energy. It means either that they traverse a more dilute material or that they travel more directly from their source to the Solar System.

Detailed calculations have been performed to show that the light elements Li, Be, B, can be produced by the action of the GCR on the interstellar medium. For these calculations, the propagation of the cosmic rays inside the interstellar medium has been sketched following a simple but reasonable approach: the galactic disk is assumed to be like a cylindrical box inside which the GCR sources are distributed homogeneously. These sources continuously produce a time independent flux of high energy particles. Three possible events can occur to these particles along their path inside the galactic disk (Figure VII.7).

(1) They approach the limit of the box and can escape from it.

(2) They encounter interstellar nuclei with which they interact by spallation reactions.

(3) They inelastically collide with the interstellar atoms and hence induce their ionization. These inelastic collisions slow down the GCR particles.

These three effects are taken into account in the transport equation of the cosmic rays (see e.g. Meneguzzi et al., 1971).

The solution of this equation leads to an exponential distribution of the mean free path for the GCR particles through the interstellar gas:

$$P(X) = \frac{1}{\Lambda} \exp\left(-\frac{X}{\lambda} \right),$$

where X is the GCR path length across the interstellar medium expressed in g cm^{-2} and Λ is the average mean free path against escape and spallation effects: at very high energy ($E > 10$ GeV/N) the ionization or electronic collision effects can be considered as negligible and Λ depends only on observable quantities such

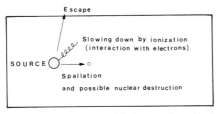

Fig. VII.7.   Sketch of the fate of the GCR particles interacting with the interstellar medium.

as the ratio of the secondary to the primary fluxes and the spallation cross sections. Let us recall that the primary elements are those which come directly from the GCR sources while the secondary ones are those produced on their way in the interstellar medium. The observed Li Be B/CNO ratio (~ 0.23) leads to $\Lambda \sim 6$ g cm^{-2}. Such path lengths correspond to a GCR lifetime of $\sim 4 \times 10^6$ yr if the interstellar density is $n_H \sim 1$ particle cm^{-3} (i.e. the average density of the interstellar medium in the galactic disk) and to a lifetime $\geq 4 \times 10^7$ yr if $n_H < 0.1$ particle cm^{-3} (average interstellar density in the galactic halo). The way to determine more accurately the 'age' of the cosmic rays, i.e. the time elapsed between their acceleration and their detection, is to look at the presence or absence of long lived radio-isotopes such as ^{10}Be (halflife $\sim 1.5 \times 10^6$ yr). Current isotopic determinations made for the GCR Be and B seem to show that a large part of ^{10}Be has decayed. Then the 'age' of the GCR is at least $\sim 2 \times 10^7$ yr which corresponds to interstellar densities <1 particle cm^{-3}.

From the cosmic ray mean free path one can evaluate the primary abundances of the cosmic rays at the source (by deducing the secondary contribution due to the interaction between the cosmic rays and the interstellar medium). Table VII.2 gives the primary abundances. One can notice the enrichment in the cosmic rays of the heavy elements relative to H and He. Nevertheless, but for the H and He abundances there is growing evidence that these GCR primary abundances are rather similar to the solar abundances. The primary GCR material could be the interstellar material surrounding supernovae (or rather active stars) accelerated by the explosion.

The light elements can be formed by the interaction cosmic rays - interstellar gas following two different mechanisms (i) the rapid (cosmic ray) H and He nuclei

TABLE VII.2

Comparison between the standard abundances (Cameron, 1973), the observed GCR composition and the primary GCR composition at the source (i.e. before any effect due to its transport through the interstellar matter occurs).

Element	GCR observed composition	Primary GCR composition (at the source)	Standard abundances
H	5.8×10^4	4.2×10^4	2.7×10^5
He	3400	3000	1.9×10^4
Li	16	$< 10^{-4}$	4×10^{-6}
Be	11	$< 10^{-4}$	7×10^{-8}
B	27	$< 10^{-4}$	3×10^{-5}
C	100	100	100
N	27	11	31
O	86	110	180
Ne	20	25	29
Mg	21	29	9
Si	15	26	8.5
Fe	11	25	7

bombard the interstellar heavy elements directly providing interstellar light elements (almost at rest), (ii) the rapid CNO nuclei bombard the interstellar H and He atoms and become Li Be B GCR nuclei. These newly formed Li, Be and B nuclei eventually slow down and contaminate the interstellar medium. These two contributions have to be summed up in order to evaluate the number of light elements produced per unit time and unit volume: with the observed GCR flux (2.5×10^{-2} protons $cm^{-2} s^{-1} GeV^{-1}$ at 10 GeV/nucleon) and with an injection spectrum (at the source) $q \propto W^{-2.6}$, where W is the total energy (rest must + kinetic energy) of the GCR, one finds that the contribution of processes (i) and (ii) is, respectively, 70% and 30% of the total formation rate.

The results of these computations are given in Table VII.3. If one integrates the present observed GCR flux over 10^{10} yr which is a rough estimate of the assumed age of the Galaxy, one sees that GCR are likely to produce the observed ^6Li, ^9Be, ^{10}B, and ^{11}B, while ^7Li is underproduced by about a factor 10 and (D and ^3He) by at least a factor 100.

These results can be slightly modified if one makes the assumption that there exist high fluxes of low energy particles which are prevented from entering the Solar System by solar modulation effects. There is yet no observation in contradiction with the existence of these fluxes at the vicinity of the cosmic ray sources: these suprathermal particles cannot go very far from their sources because of the large energy losses which they suffer. Such low energy fluxes are interesting because they can induce some ^7Li (and also ^{11}B) production according to: $^4He + {}^4He \rightarrow {}^7Li$ and $^{14}N(p, \alpha) \rightarrow {}^{11}C (\beta^+) {}^{11}B$ which are the spallation reactions with the lowest thresholds.

These low energy fluxes can explain the observed $^7Li/^6Li$ ratio which is >10 while the high energy GCR leads to a $^7Li/^6Li$ ratio <2. Observations of γ rays of a few MeV which can be produced by these low energy fluxes would either rule out or confirm this hypothesis: spallation reactions produce excited nuclei which deexcite by releasing γ rays of a few MeV.

D is underproduced (relatively to its observed abundance $D/H \sim 10^{-5}$) by factors ranging from 10 (presence of low energy fluxes) to 1000 (absence of large fluxes of

TABLE VII.3

Formation rate of light elements through the bombardment of the interstellar matter by the Galactic Cosmic Rays. One notes that this mechanism accounts rather well for the abundances of ^6Li, ^9Be, ^{10}B, and ^{11}B (Meneguzzi et al., 1971).

Light elements	GCR production rate	Abundance due to GCR (relative to H)
^6Li	1.1×10^{-4}	8×10^{-11}
^7Li	1.7×10^{-4}	1.2×10^{-10}
^9Be	2.8×10^{-5}	2×10^{-11}
^{10}B	1.2×10^{-4}	8.7×10^{-11}
^{11}B	2.8×10^{-4}	2×10^{-10}

low energy particles). It is therefore almost impossible to explain the D abundance of the Solar System and the solar neighborhood by spallation reactions induced by cosmic rays. However, in the Galactic Center, the D/H ratio can be ten times smaller than it is elsewhere. The D which has been detected there might be due to a GCR bombardment of the interstellar medium in this region (see Chapter IX).

VII.4. Light Element Production in Stellar Objects

VII.4.1. LIGHT ELEMENT PRODUCTION IN SUPERNOVAE EXPLOSIONS

Colgate (1974) has tried for a few years to build up a model in which the light elements including D are produced during the supernovae explosion themselves, inside their external layers.* The physical principle of this model is the following: when a shock wave propagates in a supernova envelope, it transports an energy which is shared between fewer and fewer particles when one goes from the internal to the external layers. When the density ρ is as low as $10^{-8} < \rho < 10^{-5}$ g cm^{-3}, the model predicts that a large fraction of the particles can reach kinetic energies higher than 30 MeV/ nucleon. This allows the complete disruption of ^{4}He into protons and neutrons. The density of these layers is such that, instead of decaying, the neutrons can recombine with the protons and produce some D during the cooling phase occurring after the passage of the shock wave. This cooling is due to the transfer of energy first to the nuclei, then to the electrons (by ionization processes similar to those discussed previously) and finally from electrons to photons by inverse Compton and bremsstrahlung mechanisms. In the inverse Compton mechanism, the high energy electrons transfer their energy by simple collision to the photons irradiating the gas. The inverse Compton process can be schematized as:

$$e^-(high\ E) + \gamma(low\ E) \rightarrow e^-(low\ E) + \gamma(high\ E).$$

The bremsstrahlung mechanism consists in an inelastic collision between electrons and atomic nuclei which ends up by the creation of photons. In other words, in this process the kinetic energy of a charged particle is transformed into radiation energy. The cooling mechanism is in fact the critical feature of the model: it should not be too rapid otherwise the maximum energy given to the nuclei by the shock wave would be too small to induce the spallation of He. It should not be too slow otherwise the deuterium which is created in the process would be spalled as well as helium.

This model is very interesting in the sense that it can provide a single explanation for the origin of all the light elements: the Li, Be and B could also be produced by spallation reactions on the CNO nuclei. However, various criticisms based on different physical arguments have been presented. The most serious problem is that of the cooling time scale. There are several cooling effects which act on the matter suffering the shock wave and which have not been included in the first versions of the model (see e.g., Weaver and Chapline, 1974).

Let us conclude by saying that the attempt to produce D within the supernovae

---

* A similar model has been proposed with supermassive star explosions. An additional shortcoming of this model is that supermassive stars are still considered as hypothetical objects.

explosions is physically interesting. Unfortunately, this mechanism cannot be used to solve the problem of the origin of the light elements.

VII.4.2. OTHER STELLAR SITES TO PRODUCE ^{7}Li

^{7}Li deserves a special mention in nuclear astrophysics; it is the only element which (to our knowledge) can be produced by more than two different nucleosynthetic processes. Contrary to the other light elements (D, ^{6}Li, Be, and B) ^{7}Li can be produced within some stars either during the Red Giant phase or by the explosive H burning which might take place during nova or super massive star outbursts.

(1) ^{7}Li production in Red Giants: In Red Giants ^{7}Li can be produced either by spallation occurring within their envelopes or by thermonuclear reactions occurring at the border of the H and He burning zones. In this last case, the ^{7}Li production goes together with that of ^{3}He for which Red Giants constitute a likely source too.

^{7}Li can be produced by spallation reactions, in particular through the (^{4}He + ^{4}He) reactions which possibly occur in the envelopes of Red Giants: contrary to the case of T Tauri stars, there is enough energy in Red Giants to allow the spallation reactions to take place due to the thermonuclear reactions inside these stars (while in T Tauri stars the energy has only a gravitational origin).

There is a thermonuclear process which can be advocated to produce ^{7}Li in red giants. A few red giant stars show an overabundance of Li: $Li/H \sim 10^{-7}$ i.e., 100 times the Solar System value. As has been seen in Chapter IV, ^{3}He can be produced in amounts such as $^3He/H \sim$ a few 10^{-5} inside the H burning zone. Red giants suffer large mass losses from which the interstellar gas can be contaminated in ^{3}He. It has also been seen in Chapter V that these stars can suffer He flashes because of the degeneracy of the He burning zones. When He flashes take place they can induce a partial mixing between the H burning zone, in which there is some ^{3}He, and the He one. The result of this mixing is that the reaction $^3He + {}^4He \rightarrow {}^7Be + \gamma$ can take place. These zones are generally subject to important convective motions which can bring ^{7}Be with ^{3}He to the surface of the red giants. In these external zones, the temperature is low enough to prevent ^{7}Be from absorbing protons. ^{7}Be can only capture electrons and be transformed into ^{7}Li with a time scale of about 50 days.*

Novae and/or supermassive stars (if these hypothetical objects do exist) can also produce ^{7}Li by explosive He burning. Too many different processes can be invoked to explain the formation of ^{7}Li: the situation is far from being clear! Nucleosynthetic studies of objects as different as red giants, novae and supernovae should be undertaken to choose between these too numerous possibilities.

VII.5. The Big Bang Nucleosynthesis

It was shown at the beginning of this monograph that there are at least two arguments to indicate that the Universe as its birth has gone through a very dense and hot phase called 'Big Bang': these are (1) the recession of the galaxies (Hubble expansion

---

* ^{7}Be, which is unstable in the laboratory, would be stable in a medium devoid of electrons.

law), and (2) the discovery of the 2.8 K black body radiation. It has been known for about 10 yr that if the Big Bang did take place, nucleosynthetic processes can have occurred during this exploding phase and synthesize elements like D, ^{3}He, ^{4}He, and ^{7}Li. Before analysing the nucleosynthetic processes in the Big Bang let us review the different cosmological assumptions which are made to build up a model of Universe expansion.

Cosmologists like Wagoner (1973), who have dealt with this problem, distinguish two different kinds of cosmological assumptions. The first two basic assumptions define any kind of Big Bang model. The other ones are more restrictive and define only the so-called Standard Big Bang. It is in the frame of this simple standard model that the nucleosynthesis of the light elements has been extensively analyzed. This simple model accounts for the observed D, ^{3}He, and ^{4}He abundances. However, it may be far from the actual evolution of the Universe.

VII.5.1. THE BASIC ASSUMPTIONS

Any Big Bang model supposes at least two assumptions: (a) the principle of equivalence is valid, i.e., the laws of physics which do not depend on gravity are the same everywhere in the Universe. This assumption is adopted in most of the gravitational theories, in particular in the General Relativity; and (b) the Universe has experienced a very dense hot phase in which all the particles were in statistical equilibrium.

VII.5.2. THE STANDARD MODEL

The Standard model is defined by five additional assumptions:

(a) There is no significant amount of antimatter in the Universe (the Universe is not symmetrical). In other words the baryonic number $B = |N_B - N_{\bar{B}}|$, where N_B is the number density of baryons (or matter) and $N_{\bar{B}}$ is the number density of antibaryons (or antimatter) is positive and large.

(b) Similarly, one can define the lepton number which is the difference between the number density of leptons (positrons, positive muons and neutrinos) and of antileptons (electrons, negative muons and antineutrinos). All these particles obey the Fermi-Dirac statistics; if they are very numerous they (in particular the neutrinos) become degenerate i.e. they do not determine the pressure of the Universe. To avoid this degeneracy, one assumes that the lepton number is much less than the number density of the photons in the Universe.

(c) The Standard model assumes that there is no elementary particle in the nucleosynthetic phases of the Big Bang other than photons, neutrinos and neutrettos (muonic neutrinos), electrons and positrons, protons and neutrons.

(d) The Universe is assumed to be homogeneous and isotropic. It satisfies the so-called 'cosmological principle' of homogeneity and isotropy.

(e) Finally, the gravitational theory of General Relativity applies to the observed Universe.

We will see later the influence of these hypotheses on this nucleosynthetic mechanism.

The main difference between the nucleosynthesis occurring during the Big Bang and the H burning studied in Chapter IV is that neutrons which are in statistical equilibrium with protons do exist in the early Universe. They can *rapidly* induce the following reaction: $n + p \rightarrow D + \gamma$.

Nucleosynthesis calculations in the Standard Big Bang have been performed by Wagoner (1973). The important characteristics of this model are that: (a) The expansion time scale of the Universe is equal to the free fall time scale defined in Chapter V:

$$\frac{1}{V(t)} \frac{dV(t)}{dt} = (24 \pi G\rho)^{1/2},$$

where V is the size of a volume element at time t. (b) The expansion is adiabatic: the relation between the density of the matter and the temperature is: $\rho = hT_9^3$ where h is also called the density parameter of the Universe model and remains constant throughout the Universe expansion. The range of possible values for the present density of the Universe is: $3 \times 10^{-32} < \rho_0 < 3 \times 10^{-29}$ g cm^{-3}.*

We find that the density parameter of the Universe should range between $10^{-6} < h < 10^{-3}$. Above 10^{10} K the neutron/proton ratio is governed by the following reactions:

$$n + e^+ \rightleftarrows p + v_e$$

$$n \rightarrow p + e^- + v_e$$

$$n + v_e \rightleftarrows p + e^-.$$

Below 10^{10} K the thermal equilibrium which existed between the particles does not hold anymore. The n/p ratio is then frozen and the following set of reactions takes place:

$$n + p \rightarrow D + \gamma$$

$$D + p \rightarrow {}^3He + \gamma$$

$${}^3He + {}^3He \rightarrow {}^4He + 2H + \gamma$$

$${}^3He + {}^4He \rightarrow {}^7Be + \gamma$$

$${}^7Be + e^- \rightarrow {}^7Li + v.$$

The only nuclear species which are produced during the Big Bang are D, ^{3}He, ^{4}He and ^{7}Li. The results of the nucleosynthesis which occurs during the Standard Big Bang are displayed in Figure VII.8 as a function of h or ρ. The important parameter is indeed the present density of the Universe which is related to the density of the Universe during the nucleosynthetic phase. For very low values of the density, the reactions able to destroy D have rather low rates while these rates increase when ρ

---

* The lower limit corresponds to the density of the visible matter and the upper limit to the critical value above which it would correspond to a too short age of the Universe.

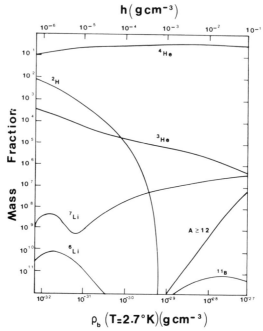

Fig. VII.8. Final abundances produced by Standard Big Bang models versus the present density of the Universe.

becomes larger. Therefore, a very important result is that D in the Standard Big Bang is produced in enough quantities to explain the observed D abundance but only if $\rho_0 < 5 \times 10^{-31}$ g cm^{-3}.

One sees the importance of D as a possible cosmological monitor. The constraint on the density, which is provided by the observed abundances of D, is to reduce the range of the possible present density of the Universe to $10^{-32} < \rho < 5 \times 10^{-31}$ g cm^{-3}.

The nucleosynthesis of D is therefore related to an important cosmological problem: as said in Chapter II the Universe which is presently expanding can be either open (i.e. expanding forever) or closed (it will contract eventually). If the present density of the Universe is $\rho_0 > 3 \times 10^{-29}$ g cm^{-3}, the matter exerts a gravitational attractive force on the Universe and compels it to pulsate (the Universe is closed). For lower densities the gravitational force never prevents continuous expansion of the Universe (the Universe is open). If D is synthetized in the Standard Big Bang the Universe should be open. In this case, another related consequence is that there should be a strict limit on the mass of invisible matter in the form of black holes, dust, low mass stars ... Let us note also that the age of the Universe deduced from the Hubble flow of the galaxies is significantly larger ($t \geq 10^{10}$ yr) in the case of an open Universe than in the case of a closed one ($t \sim 7 \times 10^9$ yr) — (cf. Chapter VIII). Therefore, the presence of D is in favor of an older Universe.

Finally, a Standard Big Bang model in which $\rho_0 = 3 \times 10^{-31}$ g cm^{-3} would explain the observed abundances of D, ^{3}He and ^{4}He and a fair fraction of ^{7}Li.

TABLE VII.4

D

He^4		Galactic origin (not produced by Big Bang)	Cosmological origin (produced by Big Bang)
	Galactic origin (not produced by Big Bang).	Degeneracy of neutrinos or symmetrical Universe (antimatter $\equiv$ matter) or General Relativity not applicable as gravitation theory.	Existence of super-baryons (Bootstrap models) or weak degeneracy of neutrinos or General Relativity not applicable as gravitation theory.
	Cosmological origin (produced by Big Bang).	Standard model with $\rho_0 > 5 \times 10^{-31}$ g cm^{-3} (closed Universe).	Standard model with $\rho_0 < 5 \times 10^{-31}$ g cm^{-3} (open Universe).

(ρ_0 present density of the Universe).

Table VII.4 summarizes the conditions which lead either to cosmological or to a galactic function of the D and ^{4}He. For example, if there is a large fraction of anti-matter, nucleosynthesis cannot proceed and there is no way to synthetize D and ^{4}He in such a Big Bang Model. If one adopts an other gravitation theory than the General Relativity, the rate of expansion and therefore the n/p ratio at the end of the equilibrium is affected. The degeneracy of neutrinos can affect also the equilibrium between the protons and the neutrons.

In conclusion, the merit of the Standard Big Bang is to explain in a natural way the observed D, ^{3}He, ^{4}He and even a fair fraction of ^{7}Li. The nucleosynthesis of the light elements put some interesting constraints on cosmological parameters such as the present density of the Universe which determines the curvature of the Universe (closed or open).

VII.6. Conclusion

The light elements hold a unique position by the fact that they are not produced in sufficient amount during the stellar evolution. Table VII.5 displays our favored nucleosynthetic sites for these elements: The Standard Big Bang appears to be a likely site for the formation of D, ^{3}He, ^{4}He and part of ^{7}Li. The GCR bombarding the interstellar medium can easily account for the formation of ^{6}Li, ^{9}Be, ^{10}B and ^{11}B and even of ^{7}Li if large fluxes of low energy (a few MeV) GCR do exist. The

TABLE VII.5

	Big Bang	GCR observed	GCR + low energy GCR	Novae	Red Giant phase
H	yes	no	no	no	no
D	yes	not enough	not enough	no	no
^{3}He	yes	not enough	not enough	possible	possible
^{4}He	yes	no	no	no	not enough
^{6}Li	no	yes	yes	no	no
^{7}Li	likely	not enough	possible	possible	possible
^{9}Be	no	yes	yes	no	no
^{10}B	no	yes	yes	no	no
^{11}B	no	not enough	yes	possible	no

case of ^{7}Li appears to be more complicated because it can be produced in various ways including novae or supernovae star explosions and nucleosynthesis during the Red Giant phase.

In any case, it is necessary to invoke at least two different nucleosynthetic processes to explain the formation of these light elements.

References

Quoted in the text:

Cameron, A. G. W.: 1973, *Space Sci. Rev.* **15**, 121.
Colgate, S. A.: 1974, *Astrophys. J.* **187**, 321.
Meneguzzi, M., Audouze, J., and Reeves, H.: 1971, *Astron. Astrophys.* **15**, 337.
Reeves, H., Fowler, W. A., and Hoyle, F.: 1970, *Nature* **226**, 727.
Wagoner, R. V.: 1973, *Astrophys. J.* **179**, 343.
Weaver, T. A. and Chapline, G. F.: 1974, *Astrophys. J. Letters* **192**, L57.

Further readings:

Audouze, J., Meneguzzi, M., and Reeves, H.: 1976, in Shen and Merker (eds.), *Spallation Nuclear Reactions and their Applications*, D. Reidel Publ. Co., Dordrecht, Holland, p. 113.
Boesgaard, A. M.: 1976, *Publ. Astron. Soc. Pacific* **88**, 353.
Reeves, H.: 1971, *Nuclear Reactions in Stellar Surfaces and their Relations with Stellar Evolution*, Gordon and Breach Science Pubs., Inc., New York.
Reeves, H.: 1974, *Ann. Rev. Astron. Astrophys.* **12**, 437.
Reeves, H., Audouze, J., Fowler, W. A., and Schramm, D. N.: 1973, *Astrophys. J.* **179**, 909.
Reeves, H. and Meyer, J. P.: 1978, *Adtrophys. J.* **226**, 613.
Schramm, D. N. and Wagoner, R. V.: 1975, *Physics Today* **27**, 40.

CHAPTER VIII

NUCLEOCHRONOLOGIES AND THE FORMATION
OF THE SOLAR SYSTEM

The endeavor of the historian is to try to date recent events which have affected the history or the evolution of humanity and the world by using mainly written records. This means that the historian cannot describe such an evolution before the advent of the writing (i.e. over a period of a few thousands years). The prehistorian who collects human or humanoid bones, artifacts such as tools, jewels, statues, potsherds of pottery . . . can describe in principle the evolution on earth for periods up to 10^6 yr. The paleontologist, who found fossils of a few 10^9 yr old algae, is able to cover eras which have almost the age of the earth itself. The task of the astrophysicist is to try to provide a chronology of the whole Universe. The main questions are: Did singular events affect the evolution of the Universe? What is the age of the celestial bodies and in particular of the Solar System? At what times have the different chemical elements been formed?

The questions regarding the nature and the evolution of the Universe, namely its possible birth (and age), are covered by cosmological studies. In fact, as we will see in the first section of this chapter, the age of the Universe can in principle be determined (of course with a rather large uncertainty) by purely astronomical techniques such as the use of the recession velocity of the galaxies (after Hubbles' discovery already mentioned in Chapter II) and also the use of techniques determining the age of the oldest stars such as those belonging to globular clusters. In the next sections, we will show how long lived radioactive nuclei (as for instance U, Th, and Pu, see a more complete list in Table VIII.1) can be used as chronometers to determine the age of the elements, i.e. the time at which the elements have been formed by the various nucleosynthetic processes. The best chronometers are indeed those which beta decay in time scales comparable to the periods one wants to determine. The composition variations due to their decay are then detectable. This implies that the chronometers should have lifetimes larger than $\sim 10^7$ yr. As can be seen from Table VIII.1, these chronometers are generally synthetized by the r process. Although they are in majority only built up by a specific nucleosynthetic process, they can be used to provide a chronology of the whole nucleosynthesis if one makes the plausible assumption that a large fraction of the nucleosynthetic processes occur almost at the same time and in the same objects. From these chronometers one can derive different interesting time scales such as the mean age of the elements, the isolation time scale of the Solar System i.e. the time between the end of the nucleosynthesis and the formation of the Solar System, the time at which the solidification of the Solar System took place . . .

Finally, we end this chapter which is not only devoted to chronologies of events

TABLE VII.1

Some important chronometers

Parent	Daughter	$\tau_{1/2}$ (yr)	Nucleosynthetic process
^{26}Al	^{26}Mg	7.4×10^5	Explosive nucleosynthesis
^{40}K	^{40}Ca	1.3×10^9	Explosive nucleosynthesis
^{87}Rb	^{87}Sr	4.7×10^{10}	r process
^{129}I	^{129}Xe	1.7×10^7	r process
^{146}Sm	^{142}Nd	10^8	p process
^{187}Re	^{187}Os	4.3×10^{10}	r process
^{232}Th	^{208}Pb	1.4×10^{10}	r process
^{235}U	^{207}Pb	7×10^8	r process
^{238}U	^{206}Pb	4.5×10^9	r process
^{244}Pu	^{232}Th fission	8.2×10^7	r process
^{247}Cm	^{235}U fission	1.3×10^7	r process

concerning the whole Universe, but also and especially to those which concern the Solar System, with a review of very recent discoveries of isotopic composition variations in some well defined fractions of meteoritic samples. As we shall see these discoveries have some implications not only on the nucleosynthetic history of the Solar System but can also provide some clues on the ways by which it has been formed.

VIII.1. Astronomical Chronologies

(a) As has been seen already in Chapter II, the Universe appears to be presently expanding. This comes from the observations of the recession velocities of the galaxies undertaken first by Hubble and more recently by Sandage and Tammann, for instance. From these observations, which constitute, with the discovery of the 3 K black body radiation, the strongest indications for the occurrence of a primordial phase in the early stages of the Universe, one knows that the relative velocity of a given galaxy increases with its distance:

$$v = Hd,$$

where H is the Hubble constant, i.e., the term of proportionality between the recession velocity and the distance of the Galaxy. Current values of H range from 50 km^{-1} s^{-1} Mpc^{-1} to about 80 km s^{-1} Mpc^{-1}. As seen from Figure VIII.1, an upper limit for the age of the Universe can be provided by the inverse of the Hubble constant

$$t < \frac{1}{H} \quad \text{i.e.} \quad t < 1.2 \text{ to } 2 \times 10^{10} \text{ yr.}$$

One should be aware here that this dating method is very uncertain (i) because one does not know accurately the deceleration factor q of the Universe, and (ii) because basically the distance of the most distant galaxies cannot be estimated precisely.

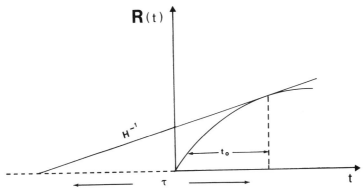

Fig. VII.1.   The age of the Universe t_0 is always less than the Hubble time $\tau = 1/H$ (which is determined by the tangent to the curve $R(t)$ i.e., the length scale between galaxies.

(b) As already stated in this monograph, the oldest observed stars are those which belong to the globular clusters. They are also called population II stars and are characterized by the fact that they belong to the so called galactic halo and have generally rather weak metal abundances. The age of the globular clusters can be estimated as for any stellar cluster by using a method designed by Sandage in 1965 and based on the magnitude length of the Main Sequence strip seen on a Hertzsprung–Russell diagram (Figure VIII.2). When the cluster is young, even the more massive stars, which burn very quickly their H,* are still on the Main Sequence and the cluster shows a rather long Main Sequence strip. On the other hand, for the oldest observed clusters the very low mass stars $M \lesssim 1\ M_\odot$ are the only ones still on the Main Sequence: These clusters show a very short Main Sequence strip. The point where the stars of a given cluster leave the Main Sequence (turn off point)

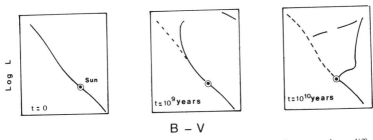

B – V

Fig. VIII.2.   Schematic representation of the HR diagram of a cluster of stars at three different epochs in its history: age zero, 10^9 yr (open cluster like the Pleiades), 10^{10} yr (globular cluster). The size of the Main Sequence (or the position of the *turn-off* point out of it) can be used as a measure of the age of the stellar clusters. This method has been proposed by Sandage and provides an estimate of about 14×10^9 yr for the globular clusters, which are the oldest objects associated with the Galaxy.

* As will be recalled in Chapter IX, the time scale during which a star with a mass of m solar masses remains on the Main Sequence is $\tau \sim (10^{10}/m^3) + 10^6$ yr.

provides a satisfactory estimate of the age of the cluster. From calculations of stellar evolution, the age of the globular clusters is $\sim 14 \pm 3 \times 10^9$ yr which is in good agreement with the age obtained from the Hubble expansion rate. Of course, the age of the Universe should be larger than the one of the globular clusters. The uncertainties regarding the age of the globular clusters are the same as those encountered in stellar evolution because the He abundance at the surface of these cool stars is not directly measurable. Therefore, it appears useful to find other independent ways to estimate the age of the Universe.

VIII.2. Nucleochronologies

The principle of nucleochronology techniques is to measure isotopic ratios in terrestrial rocks, lunar samples, planetary atmospheres ... which can be modified by the decay of a long lived radio nuclide. This is the case of the ratio of the Pb isotopes which can be affected by the decay of ^{232}Th, ^{235}U, and ^{238}U. Rutherford was indeed the first physicist in 1929 who studied the Pb isotopic ratios in various terrestrial rocks to deduce an age of solidification of these rocks. Now one generally prefers to measure the U isotopic ratios directly. The other elements which are currently used in nucleochronologies are Xe, Sr and Re: ^{129}Xe is affected by the decay of ^{129}I (half life of 1.7×10^7 yr) while ^{131}Xe, ^{132}Xe, ^{134}Xe, and ^{136}Xe are affected by the fission decay of ^{244}Pu (half life of 8.2×10^7 yr); ^{87}Rb decays into ^{87}Sr with a half life of 4.2×10^{10} yr; ^{187}Os decays into ^{187}Re with a half life of 4.3×10^{10} yr.

To estimate the age of the Solar System, the following procedure is adopted. If P is a radioactive nucleus with a decay constant $\lambda = 1/\tau$, the variation of the abundance of P (which is also denoted as P) with time is given by:

$$\frac{dP}{dt} = -\lambda P.$$

This very simple differential equation can be integrated between a given time t and the present time t_0. The origin of time is taken as the formation time of the Solar System, which means that t_0 is indeed its age:

$$P_0 = P e^{\lambda(t - t_0)}$$

where P_0 is the present abundance of the element P and P its abundance at time t.

Assume now that this radioactive element is transformed into a daughter isotope D. At any time the sum of abundances of P and D remains the same:

$$P + D = P_0 + D_0.$$

If one divides this identity by the abundance of another stable isotope of D, D_x, which receives no contribution from a radioactive nucleus, one has:

$$\frac{P}{D_x} \left| \exp\left[\lambda(t - t_0) \right] - 1 \right| + \frac{D_0}{D_x} - \frac{D}{D_x} = 0.$$

If one can measure P/D_x and D/D_x in different samples (or several separates of the same sample), one obtains in the plane P/D_x, D/D_x a straight line whose slope provides a direct determination of $(t - t_0)$.

As shown on Figure VIII.3, the slope of the line $^{87}Sr/^{86}Sr$, in function of $^{87}Rb/^{86}Sr$ gives a measure of the age of the rock for which these ratios are determined. The accuracy of the measurements performed in laboratories such as those of Institut de Physique du Globe (C.J. Allègre and his group) and Caltech (G.J. Wasserburg and his group) is quite impressive: These groups are able to determine the age of terrestrial or Solar System material with uncertainties $\lesssim 10^{-6}$ to 10^{-5}.

As developed further, the value of the Rb–Sr technique to evaluate the age of the terrestrial rocks (and therefore, the Solar System) is due to the very great stability of ^{87}Rb for which the relative ratio (compared to other Rb isotopes) does not depend much on the actual age of the Universe. There are indeed two classes of chronometers: (i) those with lifetimes larger than $\sim 10^9$ yr (for instance Th, U) which can be used to extract information on the duration of the nucleosynthesis, and (ii) those with lifetimes smaller than 10^8 yr (^{244}Pu, ^{129}I, ^{26}Al) which give some clues on the last events which occurred just before the solidification of the Solar System. This classification becomes clearer when one considers the general equation which describes the evolution with time of the abundance N_i of an element i:

$$\frac{dN_i}{dt}(r, t) = -\lambda_i N_i(r, t) + T_i(r, t),$$

where T_i is the production function of the element; Schramm and Wasserburg (1970) solved this equation in the simplified case where T_i is a linear function $T_i = P_i p(t)$ in which $p(t)$ is a function describing the nucleosynthetic efficiency in the

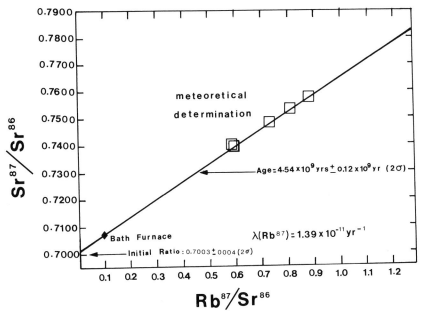

Fig. VIII.3. $^{87}Rb - ^{87}Sr$ evolution diagram for six chondritic meteorites; on the abscissa on has plotted the P/D_x (i.e. $^{87}Rb/^{86}Sr$) ratio and on the ordinate the (D/D_x) (i.e. $^{87}Sr/^{86}Sr$) ratio; the slope of the straight line provides a very accurate estimate of the solidification age of these chondrites.

considered location at a given time t. For the r process nuclei, $p(t)$ is proportional to the rate of supernova explosion responsible for this nucleosynthetic process. P_i is the relative production factor of the nucleus. The relative production ratio P_i/P_j of two nuclei i and j can be considered to be independent of time. These production factors can be directly deduced from the r process theories. It is interesting to note that in regions of the chart of the nuclides between the 'magic numbers' (see Chapter VI) the estimates of these production factors are rather easy to make: the parameters P are roughly proportional to the number of possible r process parents (or progenitors)*.

The solution of the above equation is:

$$N_i(T + \Delta) = P_i T \langle p \rangle \exp\left[-\lambda_i(\Delta + T)\right] \int_0^T \exp(\lambda_i \xi) \rho(\xi) \, d\xi.$$

In this equation T is the time elapsed between the beginning of the r process nucleosynthesis in the Galaxy history and the last nucleosynthetic event which contributed to the Solar System composition $\langle p \rangle = (1/T) \int_0^T p(\xi) \, d\xi$ is the mean rate of nucleosynthesis. Δ is the time elapsed between the end of the r process nucleosynthesis and the formation of the Solar System (solidification of planets and meteoritical parent bodies in the Solar System). Finally, the function $\rho(\xi) = p(\xi)/(T\langle p \rangle)$ represents the relative rate of nucleosynthesis compared to the average rate (Figure VIII.4).

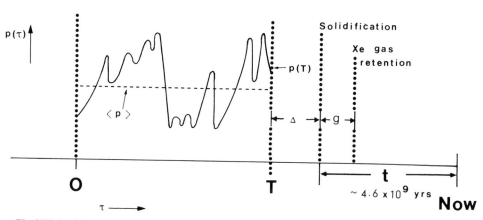

Fig. VIII.4. Scheme showing the meaning of the notation used. $p(\tau)$ is the rate of nucleosynthesis as a function of time starting at $\tau = 0$, T is the total duration of nucleosynthesis contributing to the Solar System, $\langle p \rangle$ is the average rate of nucleosynthesis and $p(T)$ is the rate just prior to the separation of the pre-Solar System gas from the galactic gas containing new nucleosynthetic ejecta. Δ represents the separation between the last nucleosynthetic event contributing to the solar system and the time of solidification of solid bodies in the Solar System, g represents the time interval between solidification and Xe gas retention (g is normally taken to be 0) and t is the age of solid bodies in the solar system (4.6×10^9 yr). The total age of the elements considered is $T + \Delta t$, which for the r process can be taken as the age of the Galaxy.

* These progenitors are the nuclei which are reached on the r process nucleosynthetic path.

If the element is stable ($\lambda = 0$) the abundance $N_i(T + \Delta)$ is simply:

$$N_i(T + \Delta) = P_i T \langle p \rangle,$$

i.e., strictly proportional to the deviation of the nucleosynthesis and its average rate. It is useful to analyze successively the long lifetime radio-nuclides such that $\lambda T \ll 1$ (corresponding to half lifetimes of at least a few 10^9 yr such as ^{238}U, ^{232}Th, ^{187}Re, for instance) from which one can determine the total duration of the r process nucleosynthesis and the short lifetime radio-nuclides such that $\lambda T \gg 1$ (lifetimes smaller than a few 10^8 yr such as ^{26}Al, ^{129}I, ^{244}Pu) which provide estimates of the time Δ between the end of the nucleosynthesis and the solidification of the solar system.

VIII.2.1. DURATION OF THE NUCLEOSYNTHESIS: LONG LIFETIME RADIONUCLIDES

A nucleochronology can be derived easily if (as shown by Schramm and Wasserburg, 1970) one introduces relative rates between a pair of two radionuclides i and j. A useful quantity is

$$R_{ij} = \frac{P_i/P_j}{N_i(\tau + \Delta)/N_j(T + \Delta)},$$

which is directly derived from the observation and studies of the r process nucleosynthesis. One can deduce another important quantity directly from the input data (this ratio and the decay lifetimes)

$$T' = \frac{\ln(R_{ij})}{\lambda_i - \lambda_j}.$$

T' is such that

$$T' = T - \langle \tau \rangle + \Delta$$

where $\langle \tau \rangle$ is the average age of the r process elements. The parameter $\langle \tau \rangle$ would represent the age of a single event which would have the same effect as the whole nucleosynthetic sequence (Figure VIII.5). The total duration of the nucleosynthesis is:

$$T = T' + \langle \tau \rangle + \Delta$$

and the 'age' of the Galaxy (as far as its nucleosynthetic properties are concerned!) is $T + t$ where t is the age of the Solar System (4.6×10^9 yr).

The average age of the r process elements depends on the evolution with time of the nucleosynthesis: if the rate of nucleosynthesis is constant, $T' = \langle \tau \rangle = T/2$; while if the rate of nucleosynthesis is a decreasing function of time $\langle \tau \rangle \to 0$ and $T \to T'$ (Figure VIII.5). Therefore, for

$$\lambda \ll 1, \quad T' \ll T \ll 2T'.$$

Pairs of nuclei which can be used for the determination of the total duration T are

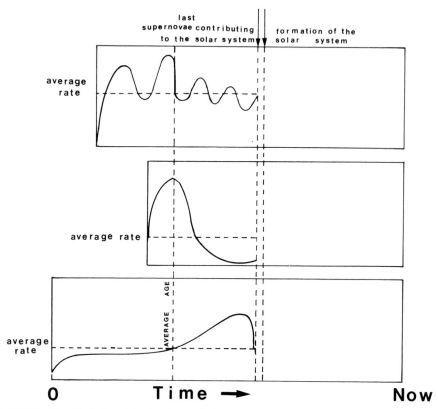

Fig. VIII.5.   Hypothetical models of nucleosynthesis postulate three possible distributions of supernovae
during the early history of the Universe. All three models assume the same 'average age' of the elements
but incorporate different 'average rates' at which the elements were formed. If the number of supernovae
has been nearly constant or if the fluctuations have been symmetrical (*top curve*), then the average age
is approximately equal to half the total age of the nucleosynthetic process. If the number of supernovae
was particularly high in an early period, (*middle curve*), then the Universe is younger than a constant-rate
model would suggest. A low early rate (*bottom curve*) yields a somewhat greater overall age. This last
model is inconsistent with observation.

given in Table VIII.2 on page 135. ^{232}Th/^{238}U is a suitable pair but difficulties may
occur from possible Th/U fractionation. The shortcoming of the ^{235}U/^{238}U pair is
the too short lifetime of ^{235}U. The best pair to be used would be ^{187}Re/^{187}Os
since ^{187}Re has a lifetime as long as $4.3 \pm 0.5 \times 10^{10}$yr. Up to a recent date the only
difficulty was that the cross sections of the ^{186}Os(n, γ)^{186}Os and ^{187}Os(n, γ)^{188}Os
were not accurately determined. Determinations are now available of the Os(n, γ)
cross sections and with these values one can deduce that T goes from 8 to about
20×10^9 yr. There are still large uncertainties even after the determination of the
neutron absorption cross sections because the ^{187}Re lifetime is not yet accurately
measured and also because the relative abundance of Os and Re could have been
changed by physical or chemical processes ('fractionation', e.g. during the solidification of meteorites).

In the frame of the current models of chemical evolution of the Galaxy (see Chapter IX) which attempt to describe the rate of star formation and death and consequently the evolution with time of the rate of supernova outbursts, the age of the r process nucleosynthesis appears to be from at least 10^9 yr up to about 20×10^9 yr.

VIII.2.2. THE SHORT LIVED ISOTOPES ($\lambda T \gg 1$)

As said before, the short lived isotopes teach the astrophysicists and the cosmochemists about the last nucleosynthetic events. The problem can be summarized as "when the parents are gone, one should try and find the children". The general equation which gives the evolution with time of the abundance of the radio nuclides can be written in this case as:

$$N_i(T + \Delta) = \frac{P_i}{\lambda_i} p(T) \exp(-\lambda_i \Delta),$$

from which one can deduce

$$\Delta = \frac{1}{(\lambda_i - \lambda_j)} \ln\left(R_{i,j} \frac{\lambda_j}{\lambda_i} \right).$$

There are two sets of nucleochronology experiments which lead to interesting determinations of the time during which the solar system isolates itself from the nucleosynthetic sources: (i) those based on the determination of Xe isotopic composition, and (ii) those based on the very recent discovery of Mg anomalies related to the decay of ^{26}Al in Al rich chondrules belonging to the meteorite called Allende which recently fell in Mexico (see below, Section VIII.3).

Several Xe isotopes are the 'children' of short lifetime radio nuclides. ^{129}Xe comes from the ^{129}I decay while the fission of ^{244}Pu produces overabundances in ^{132}Xe, ^{134}Xe, and ^{136}Xe. Since 1959, J. H. Reynolds and his associates have thoroughly studied this Xe isotopic composition in various meteorites.

(a) Figure VIII.6 shows a Xe spectrum taken from a meteorite called Richardton. The horizontal lines correspond to the isotopic composition of the atmosphere. All the meteorites show this conspicuous deviation for the mass 129. After a neutron irradiation made in the laboratory which transforms the ^{127}I of the meteoritic sample into ^{128}I (therefore ^{128}Xe), one can deduce an initial ^{129}I/^{127}I ratio of about 10^{-4} corresponding to a solar system formation time scale of about 1 to 2×10^8 yr. It is interesting to note that differences of initial ^{129}I/^{127}I ratios in various meteorites show that the various meteorites have been solidified within 14×10^6 yr. This result is referred to as the existence of a 'sharp isochronism' in the formation of the meteorites. Interesting studies of 'Xenology' showed that the various crystals of a *given* meteorite have ages as widely spread as the whole sample of meteorites themselves. This means that meteorites are a complex collection of minerals which have different solidification histories.

(b) It has been shown that the ^{130}Xe, ^{132}Xe, ^{134}Xe ratios allow the deduction of ^{244}Pu/^{238}U initial ratios. As shown in Figure VIII.7, there is a striking agreement between the isotopic pattern for Xe from spontaneous fission of ^{244}Pu and the

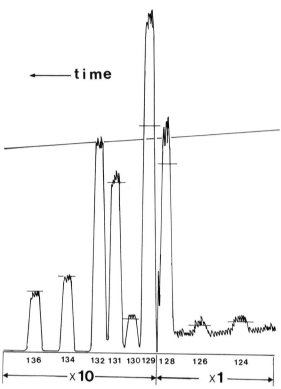

Fig. VIII.6. Mass spectrum for Xe extracted from the Richardton chondrite. The short horizontal lines mark where the peaks would fall if the sample were Xe from the atmosphere. By far the most marked anomaly is an excess at mass 129 due to radioactive decay of extinct ^{129}I.

fissiogenic Xe found in meteorites. On the contrary, one can notice that the isotopic pattern which comes from the fission of ^{244}Cm does not fit the meteoritic isotopic pattern at all.

The formation time scale deduced from the ^{244}Pu fission Xe spectrum is comparable to that deduced from the $^{129}/^{129}$Xe pair. This age of about 10^8 yr has been interpreted by different astrophysicists (e.g. H. Reeves) as corresponding to the interaction of the young solar system with a spiral arm of the Galaxy, because a time of 10^8 yr is the typical time elapsed between the passages of two spiral arms in the Galaxy. Since the density waves responsible for the spiral arms of the Galaxy are also supposed to be able to trigger a star formation process, it appears quite natural to associate these two types of events (rotation of the spiral arm pattern and Solar System formation). One should note at this point that other chronometers such as ^{146}Sm (which is a p process element produced by nucleosynthetic mechanisms related to supernovae explosions) lead to comparable solidification time scales.

The primordial abundance of ^{244}Pu allows one to also examine an important nucleosynthetic problem. All the current models which describe the chemical evolution of galaxies predict a production rate for the r process rapidly decreasing

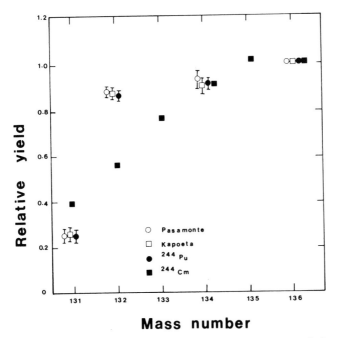

Mass number

Fig. VIII.7.   Agreement between the isotopic pattern for Xe from spontaneous fission of ^{244}Pu and fissiogenic Xe found in achondrites. By this comparison, the existence of ^{244}Pu as a *bona fide* extinct radioactivity was confirmed. Work of Alexander *et al.* (1971).

with time. However, if as we will see below, the formation of the solar system has been triggered by one or several supernovae explosions, one might expect a spike in the production rate curve just before the isolation of the Solar System (see Figure VIII.8).

By looking at one given mineral separate of a meteorite fallen in France in 1967, Saint Severin, a group of cosmochemists found ^{244}Pu$/^{238}$U $= 0.035$ (at the formation of the Solar System) while when the analysis was performed later on the whole meteoritical material ^{244}Pu$/^{238}$U was found to be significantly lower (^{244}Pu$/^{238}$U $= 0.015$). The discovery of this discrepancy has two important consequences: (i) there is no need for a spike in the r process rate curve; (ii) fractionation effects between elements can induce large errors especially when one uses pairs in which the chemical properties of the members are different like Pu and U in this case.

As will be developed in the next section, one can now make use of a very short lived chronometer ^{26}Al ($\tau \simeq 7.4 \times 10^5$ yr) which, besides its very short lifetime, presents the advantage of being produced by nucleosynthetic processes other than the r process. From measurements first undertaken by Gray and Compston (1974) and then more successfully pursued by Wasserburg and his group (1976 ...) on Al rich inclusions of the Allende meteorite, one can deduce that the primordial ratio ^{26}Al$/^{27}$Al was as high as 5×10^{-5}. As is seen in Figure VIII.9 an excess of ^{26}Mg appears to be proportional to the Al content of the sample and can be related to the decay of ^{26}Al. This measurement shows that this ^{26}Al was formed by a

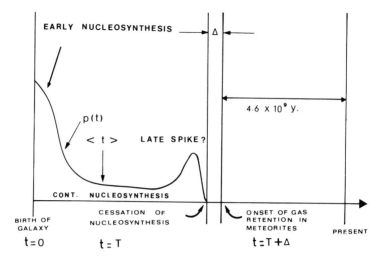

Fig. VIII.8. Scheme of the variation with time of the rate of the *r* process showing the possibility of a late spike to account for the products of the observed ^{244}Pu. Recent determinations of ^{244}Pu in Saint Severin cast doubts on the occurrence of such a late spike.

supernova a few million years (and not 10^8 yr!) before the solidification of the grains where the ^{26}Mg anomaly was found. The implications of the difference between these two formation time scales (10^8 yr from Xe and 10^6 yr from Al–Mg) will be presented in the next section. This discovery forces those involved in writing the solar system history to present a slightly complicated but hopefully convincing scenario to describe it.

To summarize what can be learned from the use of conventional (and now less conventional) chronometers: (i) the age of the Solar System is known with an ex-

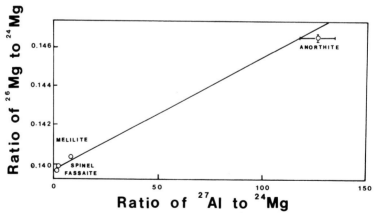

Fig. VIII.9. Excess of ^{26}Mg in an inclusion of the Allende meteorite (from Lee *et al.*, 1976). Melilite, spinel, fassaite and anhortite designate mineral separates of this inclusion, the last one being the richest in Al. The isotopic enrichment is due to the decay of ^{26}Al and is proportional to the Al/Mg element ratio. From the slope of the line one deduces that the primordial ^{26}Al ^{29}Al ratio in this inclusion was equal to 5×10^{-4}, 4.6×10^9 yr ago.

cellent accuracy thanks to the Rb–Sr techniques; (ii) the age of the r process is between 8 to 20 $\times$ 10^9 yr depending mainly on the choice of the models for chemical evolution (or stellar formation and death) of the Galaxy. Models where the bulk of the nucleosynthesis is made on the early stages of the Galaxy would favor a short age while models in which there are more continuous nucleosynthetic effects would favor longer ages (the same effect is also observed in models which allow a continuous admixture in the galactic disk of unprocessed material coming from external regions of the Galaxy); (iii) the formation time scale Δ of the Solar System does *not* lie between 10^6 to 10^8 yr: the r process chronometers (and also a p process chronometer ^{146}Sm*), observed in many different meteoritical samples, provide a consistent value for Δ of about 10^8 yr. In the case of ^{26}Al, which is observed only in a few Al–Ca inclusions of a given meteorite (Allende) and which has a different nucleosynthetic history possibly not related to the same objects responsible for the r and p process, one deduces a shorter value (10^6 yr) for the formation time scale of the Solar System; and (iv) there is apparently no need for a spike in the r process rate just before the solidification of the Solar System.

Progress which can be achieved in that type of research is:

(i) The use of different other chronometers – a p process element ^{146}Sm ($\tau = 7 \times 10^7$ yr) has already been quoted. Other short lived isotopes have also been considered: ^{247}Cm ($\tau = 1.6 \times 10^7$ yr) decays into U instead of Xe and could provide some insight on the possible fractionation between Pu and U or on the uncertainties on the determination of the separation time Δ due to the Xe volatility. Regarding the s process which should occur during the normal course of the stellar evolution and not during the supernova explosions. There are two possible chronometers. The first one, ^{176}Lu, is interesting because of its long lifetime (2.6×10^{10} yr). Unfortunately it is almost impossible to evaluate accurately its formation rate because of intrinsic uncertainties on the nuclear parameters involving this nucleus. The second one, ^{205}Pb (1.5×10^7 yr), is a more promising shortlived isotope. For this case too, there are a lot of nuclear physics experiments or calculations to be made before using it as a chronometer. However, the separation time Δ for the s process does not look (in first approximation) very different from the separation for the r process.

(ii) Much experimental and theoretical work in nuclear astrophysics has to be conducted especially concerning the determination of neutron absorption cross sections, characteristics of the nuclear excited states and life time of the unstable nuclei.

(iii) The fractionation processes which occurred during the solidification of the Solar System have to be understood. In particular, one should pursue the analysis of the chemistry which occurred then and which is obviously related to the thermal evolution of the young Solar System.

(iv) It is obviously necessary to get better insights into the nucleosynthetic processes responsible for the formation of these radionuclides and the evolution with time of the nucleosynthesis (see Chapter IX).

---

* As shown by Audouze and Schramm (1972) this derivation applied to a p process element is less straightforward than for a r process element because of the large uncertainties on the nucleosynthesis of the p process itself.

There might be some nucleosynthetic reasons responsible for the difference between the separation ages deduced from the r process and ^{26}Al (see below).

VIII.3. Isotopic Anomalies in Carbonaceous Chondrites

It is well known that the abundances of the different chemical species vary largely within the Solar System sample: for instance, H and He, which are very abundant at the surface of the Sun and Jupiter have largely escaped out of the lunar and terrestrial material. By contrast, and if one excepts a few noble gases such as He, Ne, or Xe*, the isotopic composition of the chemical elements was assumed to be remarkably uniform within the different constituents of the Solar System (terrestrial and lunar rocks, meteorites, solar and planetary surfaces . . .). Since about 1969, a growing number of experimental works have shown the existence of noticeable isotopic anomalies in mineral phases of carbonaceous chondrites. The chondrites which constitute about 2/3 of the discovered meteorites are characterized by the presence of chondrules which are millimetric or centimetric mineralogical assemblages with a more or less spherical shape. A subset of the chondrites consists of the so called carbonaceous chondrites: in this category the chondrules are embedded in a dark substratum enriched in C. This substratum often includes refractory white inclusions enriched in Ca–Al. As it has been said in previous chapters (see e.g. Chapters II and III) the carbonaceous chondrites are the most primordial components of the Solar System: they have a chemical composition quite similar to that of the Sun** (see Figure VIII.10). Therefore, they have been used together with the solar spectra to provide the table of the cosmic abundances. The mineralogical aspect of the meteorites results from the cooling sequences experienced by the Solar System during its formation: the metallic oxides and the refractory silicates with a large content of Ca and Al constitute the white inclusions and have been solidified at temperatures ~ 1450 to 1700 K; the silicates included in the chondrules and in the substratum (also called the matrix) solidify at $T \sim 1200$ to 1400 K while the carbonaceous compounds condense when the temperature is below 700 K. The fact that the carbonaceous chondrites have the higher content in volatile components is in favor of their being considered as the most primitive structures of the Solar System.

The history of the discoveries of most of the isotopic anomalies is dominated by the study of the carbonaceous chondrite Allende which fell in Pueblo Allende, Mexico 1969. Three reasons explain the major role played by this single meteorite in this fast growing field of modern cosmochemistry: (i) The Allende meteorite is indeed the largest carbonaceous chondrite which fell on the earth surface (the original body should have had a mass of about two tons); a lot of material is therefore

---

* These exceptions can be easily understood: (1) In the case of He and Ne the overabundances of ^{3}He and ^{21}Ne observed in meteorites are related to the bombardment by solar cosmic rays suffered by these rocks during the 10^6 yr period between the break-up of their parent bodies up to their fall on Earth. (2) In the case of Xe (as has been seen in the previous sections of this chapter) the overabundance of ^{129}Xe noticed in some meteorites (compared to its terrestrial value) is due to the decay of the radioactive ^{129}I nucleus while the overabundances in ^{134}Xe and ^{136}Xe are due to the decay of ^{244}Pu.
** With the obvious exception of the volatile elements H, He, C, N, O and the noble gases.

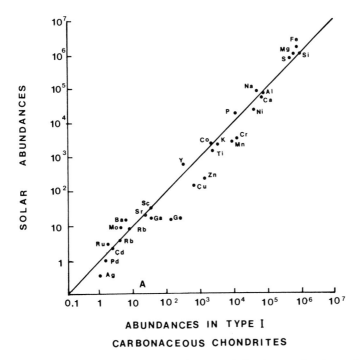

Fig. VIII.10.   Comparison between the solar abundances and the element abundances in the type I carbonaceous chondrites (normalization Si $= 10^6$).

available for cosmochemistry analysis. (ii) It is especially rich in various types of refractory mineralogical inclusions which are quite easy to separate. (iii) Many laboratories have devoted a lot of effort to refine their experimental procedures to study the lunar samples after the US (Apollo) and Soviet programs of lunar rock gathering. The improvement of experimental techniques such as those achieved in laboratories like the Geochemistry Department in Caltech 'Lunatic Asylum', the Geochemistry Department of the University of Chicago, the MacDonnell Center of Washington University, the Laboratoire Rene Bernas and the Geochemistry Laboratory of the Institut de Physique du Globe of the Paris University, has been highly beneficial to these discoveries.

A few 'isotopic anomalies', especially the large ones concerning the noble gases He, Ne, Kr, and Xe, were discovered long before the extensive analyses of Allende. We already reviewed those concerning Xe. We will only present here three other important anomalies concerning these elements:

(a) Some meteorites, called the gas rich meteorites which belong to the general class of chondrites, show an especially large relative content in ^{3}He and ^{21}Ne. These high contents in ^{3}He and ^{21}Ne are due to the spallation reactions (see Chapter VII) in the meteoritical atoms bombarded by the solar cosmic rays. The values of the ^{3}He/^{4}He and ^{21}Ne/^{20}Ne (corrected from the primordial component) provide an estimate of the time during which the meteorites were subject to this bombardment (the average solar cosmic ray flux has been uniform for at least 10^9 yr). The meteorites come from the break up of bigger objects called parent bodies. From this method one can conclude that this break up occurred a few million years ago.

(b) Ne has three stable isotopes: ^{20}Ne, ^{21}Ne and ^{22}Ne. It has been found that the terrestrial isotopic ratios are quite different from those determined from the solar wind

$$\left(\frac{^{20}\text{Ne}}{^{22}\text{Ne}}\right)_{\text{terrestrial}} = 8.2 \quad \text{while} \quad \left(\frac{^{20}\text{Ne}}{^{22}\text{Ne}}\right)_{\text{solar wind}} = 12.5$$

$$\left(\frac{^{21}\text{Ne}}{^{20}\text{Ne}}\right)_{\text{terrestrial}} = 0.025 \quad \text{while} \quad \left(\frac{^{21}\text{Ne}}{^{20}\text{Ne}}\right)_{\text{solar wind}} = 0.036.$$

In general, the Ne isotopic ratios determined in meteorites lie between these two extreme compositions. However, in a few carbonaceous chondrites like the one which fell near Orgueil (France), a third component called Ne E has been found in which ^{20}Ne/^{22}Ne is as low as 1.5 (^{21}Ne/^{20}Ne = 0.015). This material might have undergone huge irradiations by energetic particles (see Figure VIII.11).

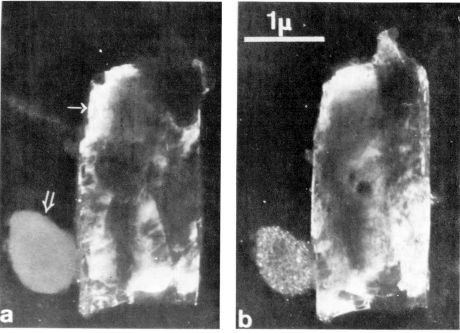

Fig. VIII.11. Dark-field microphotographs of grains from the Ne–E separate of Orgueil. (a) Before heating the grains are either crystaline (single arrow) or completely amorphous (double arrow). (b) After heating at a low temperature ($\sim 500^\circ$C) the crystallites (random crystallization) are only observed in the amorphous grains. The crystalline grains show this pattern only when they have been artificially irradiated with heavy ions and then heated at $\sim 500^\circ$C.

An explanation of this high content in ^{22}Ne was proposed by Audouze *et al.* (1976) who argued that the rather large enrichment in ^{22}Ne might also be due to spallation processes able to trigger the ^{25}Mg(p,α) ^{22}Na(β^+) ^{22}Ne (remember that the silicates which are the main constituents of these mineralogical phases have a high Mg content).

(c) In the case of Xe, quite an abnormal isotopic distribution has been found in a few specific meteoritic mineral phases. Both the light isotopes ^{124}Xe and ^{126}Xe (synthetized by the *p* process nucleosynthesis) and the heavy isotopes ^{134}Xe and ^{136}Xe (synthetized by the *r* process nucleosythesis) can be enriched with respect to ^{128}Xe by factors up to two. Two hypotheses presently attempt to explain this Xe 'special anomaly'; the first was proposed by Anders and co-workers of the University of Chicago who argued that this anomalous pattern is due to the fission spectrum of a superheavy element ($Z \simeq 110$). Other workers, like Manuel and Sabu of the University of Missouri at Rolla, prefer to interpret this pattern by specific *p* process or *r* process nucleosynthesis induced by the nearby supernovae which might have triggered the formation of the Solar System.

The anomalies concerning noble gases can be larger than a factor of two and their discovery was made before the 'Allende Era'. Now, anomalies have been found for other elements as different as O, Mg, Si, K, Ca, Sr, Ba, Nd, Sm, and the list is obviously not closed. The pioneering work of R. N. Clayton and associates of the University of Chicago on O anomalies, performed in 1973, has undoubtedly catalyzed this burst of important experiments in cosmochemistry. They found indeed a noticeable anomaly of O by mass spectrometry studies of white inclusions of Allende. The physico-chemical processes are able to produce small but significant isotopic fractionation effects which are proportional to the mass difference between the isotopes: if the fractionation increases the ratio ^{17}O/^{16}O by a quantity a, it should increase the ratio ^{18}O/^{16}O by a quantity $2a$. This is true for all the terrestrial and lunar rocks and for the meteorites other than the carbonaceous chondrites (Figure VIII.12). In white inclusions as well as in some silicate crystals of Allende and a few other

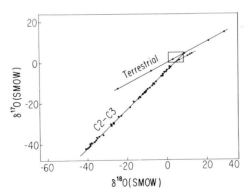

Fig. VIII.12.   Isotope plots illustrating large scale variations in O isotopic abundances. (1) A mass fractionation line defined by several terrestrial materials. (2) A mixing line defined by anhydrous phases of C_2 and C_3 meteorites, resulting from admixture of an ^{16}O rich component. SMOW = Standard Mean Ocean Water.

carbonaceous chondrites, the variations of the $^{18}O/^{16}O$ noted by:

$$\delta^{18} = \left| \left(\frac{^{18}O}{^{16}O} \right)_{\text{sample}} - \left(\frac{^{18}O}{^{16}O} \right)_{\text{standard}} \right| \frac{1000}{^{18}O/^{16}O_{\text{SMOW}}}$$

are equal to $^{17}\delta$ ($^{17}\delta$ being the same quantity for ^{17}O): the slope of the isotope variations $^{17}\delta/^{18}\delta$ is 1.0 instead of 0.5. The explanation proposed by R. N. Clayton and associates is that the anomaly comes from the admixture of 5% of pure ^{16}O to a 'normal' mixture of O isotopes. This might be understood if the grains were formed in the vicinity or in the external layers of a supernova which is more likely to synthetize ^{16}O in a He burning zone than ^{18}O and of course ^{17}O (which is only formed in H burning zones).

As has been already noted in the previous section, another very important anomaly was determined in Ca–Al rich inclusions of Allende. It concerns Mg: one notices an enhancement of ^{26}Mg (which should originate from the decay of radiogenic ^{26}Al) proportional to the ^{27}Al content ($^{26}Al/^{27}Al$) $\sim 5 \times 10^{-4}$ (Figure VIII.9). This anomaly has a similar magnitude as that of ^{16}O. It carries other important information the time elapsed between the nucleosynthetic event which produced ^{26}Al and the time of its admixture in a meteorite grain which should be of the order of 10^{6} yr. As well as ^{16}O, ^{21}Al should be produced in the explosion of a supernova but there is still presently a controversy on the actual nucleosynthetic process responsible for its formation (although explosive He burning seems to be a more likely candidate than explosive C burning).

The significant but more modest (0.01 to 1%) anomalies concerning Si, Ca, Sr, Ba, Nd, Sm ... seriously question the significance of the Standard Abundance Distribution deduced from previous analyses of the solar and the meteoritical abundances (see Chapter II). One of the most important tasks in Nuclear Astrophysics (developed in the previous chapters) was to try to explain this standard distribution. It clearly appears that the problem is far more complex since one must eventually find reasonable nuclear and astrophysical explanations for all these anomalies.

Let us restrict the discussion to the case of the O and Mg anomalies which constitute the most significant findings of modern cosmochemistry and lead to interesting ideas concerning the mechanisms by which the Solar System might have been formed.

VIII.4. The Astrophysical Implications of the Anomalies

Three different types of astrophysical implications deserve to be considered now:

(1) Various authors, including Cameron, Truran, and Schramm, suggest that the explosion of a nearby supernova has triggered the collapse of the interstellar cloud which generated the Solar System. Figure VIII.13 sketches this type of scenario in which the shock wave induced by the supernova explosion compresses the interstellar cloud and therefore facilitates the star formation. It also induces the admixture of some grains (the 'anomalous' grains) produced in the vicinity (in the external zones) of the supernova with the material parent of the solar system. This type of theory

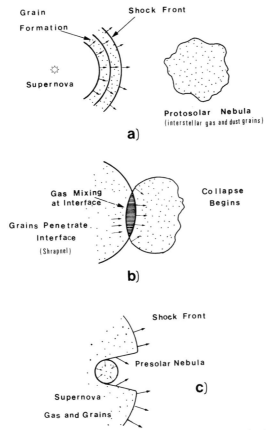

Fig. VIII.13. This scenario for the formation of the Solar System shows the four components of pre-Solar System material assumed in this paper: interstellar gas and dust of the protosolar nebula, and super-nova gas and dust coming possibly from the expanding supernova ejecta. This ejecta encounters the protosolar nebula and triggers its collapse. (b) Supernova gas mixes with nebula gas at the collision inter-faces; supernova grains may penetrate into the protosolar cloud. The collapse of the protosolar nebula proceeds in (c). The bulk of the supernova ejecta which still expands, passes around the protosolar.

predicts that the distance between the supernova and the collapsing cloud should be about 10 to 20 parsecs.

(2) Reeves (1978) proposes a scenario in which there is not only one triggering supernova but fireworks of ten or more supernovae (all coming from short lived O stars) in a very small volume: IR and radio observations regarding Orion show that the star formation in this region should occur in a very small volume (Figure VIII.14). It supports the present idea that in an O and B star association the star formation processes take place indeed in a very small fraction ($\sim 10^{-7}$) of the galactic volume. Moreover, the O stars, which have very short lifetimes (10^6 yr), can explode before leaving their formation region and contaminate the whole region of star formation.

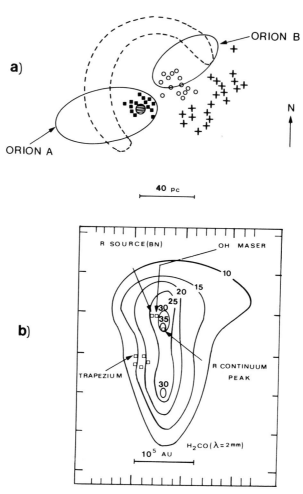

Fig. VIII.14. *The Orion OB association*: On figure (a) stars of subgroup a (Northwest) are marked by (+), subgroup b (Belt) by (○), and subgroup c (Outer Sword) by (■). The Orion nebula, the formaldehyde cloud and the IR cluster are all within the hatched area, enlarged in (b). Orion A and B are two large CO clouds. The dotted semi-ring is the Loop of Barnard. The scale is shown at the bottom. Figure (b) shows the profile of a very dense cloud (formaldehyde cloud) which contains very young stellar objects and is very near (within 10^5 AU) of the four brightest members of the Trapezium cluster.

These two scenarios, based on the basic role of one or several supernovae acting in the vicinity of the protosolar cloud, are not really new ideas: it was already developed by Opik (1953). However the discoveries of isotopic anomalies, which may have been carried into the Solar System by grains inside or in the vicinity of this (these) supernova(e), support this likely hypothesis. However, this type of model should be able to explain the difference of separation time deduced from ^{26}Al ($\sim 10^6$ yr) and from Xe isotopes ($\sim 10^8$ yr). For instance, it could be that O and B stars are not likely contributors of r process nucleosynthesis while they can contribute to the explosive He burning. In this case, only the supernovae triggered by the spiral waves of the Galaxy (periodicity of $\sim 10^8$ yr) would contribute to the r process.

(3) Finally for D. D. Clayton the anomalous grains have an interstellar origin and are in fact physically or chemically fractionated in the interstellar matter. For this author, there is no need for any nearby supernova. Supernovae contribute indeed to various anomalies but the anomalous grains may not have been locally formed.

There is no basic discrepancy between this hypothesis and the previous ones although Clayton could claim that this model is less restrictive than the two previous ones. However, in this case the short lived chronometers which can be carried by anomalous grains with unknown origin cannot be used to derive any separation time scale. Although this hypothesis seems at first more general than the other ones, there are still some difficulties based on (i) the large observational evidence that stars are formed in very restricted locations and therefore that the Solar System formation should have been influenced by many neighboring supernovae and (ii) there are too many free parameters to account for the observed anomalies: one can mix as many different anomalous grains as one wants in any arbitrary proportion to account for the isotopic anomalies.

VIII.5. Conclusion

It is very important to determine the time scales for the formation of galaxies, stars, the Solar System ... It is at first glance quite satisfactory to obtain similar time scales (around 10^{10} yr) from the relative velocities of galaxies from the HR diagram of globular clusters, and from the long lived chronometers (U, Th, Re ...). It should be realized, however that long lived chronometers give only an extremely rough estimate of this r process time scale, especially when one has to consider the large variety of models of chemical evolution of galaxies which will be discussed in the next chapter. By contrast, one knows with a luxurious number of significant digits the age of the terrestrial, lunar and meteoritical rocks from the Rb–Sr techniques.

The formation of the solar system itself is dominated by the separation time scale (related to the short lived chronometers) and by the recent but exceedingly important discoveries of isotope anomalies in carbonaceous chondrites. The role of one or several nearby supernovae at the birth of the Sun should be seriously considered.

TABLE VIII.2

Chronometer pair	P_i/P_j calculated production ratio	$T - \langle \tau \rangle = T'$ Age of the single event prior to the formation of the Solar System in 10^9 yr.	$\dfrac{N_i(T + \Delta)}{N_j(T + \Delta)}$ Abundance ratio 4.6×10^9 years ago.
$^{232}Th/^{238}U$	1.9	2.4(1 to 5.2)	2.4
$^{235}U/^{238}U$	1.5	2.0(1.3 to 2.4)	0.3
$^{244}Pu/^{232}Th$	0.5	0.5(0.4 to 0.6)	6×10^{-3}
$^{129}I/^{127}I$	1.5	0.22(0.21 to 2.5)	10^{-4}

Finally, for a nuclear astrophysicist the importance of the cosmochemistry appears to grow continuously: the Standard Table of Abundances is mainly based on the meteoritical composition. The cosmogonic problem of the formation of our Solar System may only find its solution when one solves the quite complex puzzle of all these anomalies.

References

Quoted in the text:

Alexander, E. C., Lewis, R. S., Reynolds, J. H., and Michel, M. D.: 1971, *Science* **172**, 837.
Audouze, J., Bibring, J. P., Dran, J. C., Maurette, M., and Walker, R. M.: 1976, *Astrophys. J. Letters* **206**, L185.
Audouze, J. and Schramm, D. N.: 1972, *Nature* **237**, 447.
Cameron, A. G. W. and Truran, J. W.: 1977, *Icarus* **30**, 447.
Clayton, D. D.: 1978, *The Moon and the Planets* **19**, 109.
Clayton, R. N., Grossman, L., and Mayeda, T. K.: 1973, *Science* **182**, 485.
Clayton, R. N. and Mayeda, T. K.: 1977, *Geophys. Rev. Letters* **4**, 295.
Lee, T., Papanastassiou, D. A., and Wasserburg, G. J.: 1976, *Geophys. Rev. Letters* **3**, 109.
Opik, E.: 1953, *Irish Astron. J.* **2**, 219.
Reeves, H.: 1979, *The Moon and the Planets*, to be published.
Sandage, A.: 1957, *Astrophys. J.* **125**, 435.
Schramm, D. N.: 1974, *Ann. Rev. Astron. Astrophys.* **12**, 383.
Schramm, D. N.: 1973, *Space Sci. Rev.* **15**, 51.
Schramm, D. N. and Wasserburg, G. J.: 1970, *Astrophys. J.* **162**, 57.
Wetherill, G. W.: 1975, *Ann. Rev. Nucl. Sci.* **25**, 282.

Further Readings:

Audouze, J.: 1973, in Proceedings of the European Physical Soc. Meeting, York, to be published in 'Trends in Physics', 1978, The Institute of Physics, England.
Allègre, C. J. and Michard, G.: 1973, *Introduction à la géochimie*, Presses Universitaire de France, Paris.
Fowler, W. A.: 1972, in F. Reines (ed.), *Cosmology Fission and Other Matters, a Memorial to George Gamow*, Hilger, London.
Gray, C. M. and Compston, W.: 1974, *Nature* **251**, 495.
Mason, B. (ed.): 1971, *Elemental Abundances in Meteorites*, Gordon and Breach Science Pubs., Inc., New York.
Podosek, F. A.: 1978, *Ann. Rev. Astron. Astrophys.* **16**, 293.
Reeves, H. (ed.): 1972, *On the Origin of the Solar System*, Symposium CNRS, Editions CNRS, Paris.
Schramm, D. N.: 1979, *The Moon and the Planets*, to be published.

CHAPTER IX

CHEMICAL EVOLUTION OF GALAXIES

The various nucleosynthetic processes needed to account for the formation of the different nuclear species have been described in the previous chapters as well as the location of these nucleosynthetic events. At the beginning of this monograph, the evolution of the matter which constitutes the celestial bodies such as planets, stars and galaxies was briefly recalled. During this evolution, which lasted about 10^{10} yr, the nucleosynthesis has modified the distribution of the element abundances. This chapter deals with the evolution with time of these abundances. The basic astronomical object which should be considered for the study of such time dependent evolution is the Galaxy: each galaxy is not only made of stars in which nucleosynthesis takes place and which provide the main source of galactic energy but also of interstellar gas which is contaminated by the output of the nucleosynthetic processes and from which successive generations of stars are formed.

The review of this chemical or nuclear evolution begins with a summary of the observations which should be explained by this evolution (Section IX.1). A few parameters such as the nucleosynthetic properties of stars according to their mass, the rate of star formation, the mass distribution of stars, the possible mixing of interstellar gas with fresh unprocessed material are presented in Section IX.2. Section IX.3 describes a 'simple' model tentatively designed to account for the evolution of the so-called solar neighborhood i.e. a region surrounding the Solar System. Finally, in Section IX.4, we conclude this review by mentioning some ways of improving the 'simple' model and also describing the evolution of other astronomical regions such as the center of our Galaxy and of other galaxies.

IX.1. Observational Abundance Distribution

Some of the observational facts which are the basis of any evolution model have already been mentioned at the beginning of this monograph. They are (1) the temporal and spatial uniformity of the He abundance; (2) the presence of D in relatively large amounts (although it is completely destroyed inside the stars); and (3) the temporal and spatial variation of the abundances of the heavy elements ($A \geq 12$). These temporal and spatial variations are different from one element to the other.

(1) *The* D *and* He *abundances*: As noted in Chapter VII, D and He are now believed to be formed during the first stage of the Universe (Big Bang nucleosynthesis). One should notice that their abundances do not vary much from one place to another: the canonical i.e. generally accepted D/H and ^{4}He/H values are, respectively, $\simeq 2 \times 10^{-5}$ and 0.10. However, D seems to have an abundance roughly ten times lower in the Galactic Center region while the He abundance would be somewhat

weaker in the satellites of our Galaxy (the Small and the Large Magellanic Clouds)–
(Figure IX.1). The abundance of the heavy elements is also lower in the Magellanic
Clouds than in the disk of our Galaxy: a relative underabundance in He appears
to be correlated with an underabundance in heavy elements. It is easy to understand that during the galactic evolution the He abundance should increase with
the heavy element abundances while D, which is destroyed in stellar interiors,
should see its abundance decrease in regions of active star formation (like the Galactic Center).

(2) *Heavy element abundances*: At the surface of old stars (population II or halo
stars) the abundances of the heavy elements are much smaller than at the surface
of young disk stars (population I stars). The deficiency relative to H can be as low
as 10^{-2} compared to the solar value. The abundances observed at the surface of
the stars are generally representative of the abundances of the interstellar gas from
which they have been formed. The variation of the metallicity (abundances
of the heavy elements) is shown in Figure IX.2. Nobody has found yet a single star
with a metallicity equal to zero except white dwarfs, which is a special case (metals
have sunk due to gravitational settling). The metallicity rapidly increases during
the first 10^9 yr from the halo abundances (population II) to the old disk abundances
(population I). The number of stars with low Z abundances is much less than the
number of stars showing roughly a solar abundance of heavy elements. The variation
with time of the abundances of heavy elements is not the same from one element
to the other: As we will see later on, it appears that elements which can be produced
in quiet phases of the stellar evolution (quiet nucleosynthesis in red giant phases

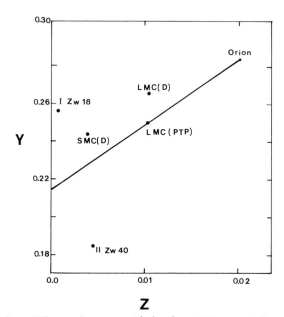

Fig. IX.1. He abundance Y (by mass) versus metal abundance Z (by mass). The points correspond to
Orion, the Large and the Small Magellanic Clouds, and two blue compact galaxies (I Zwicky 18 and II
Zwicky 40). The straight lines corresponds to $\Delta Y/\Delta Z = 3.3$.

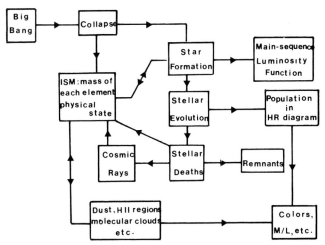

Fig. IX.2. Flow chart, as a guide to the relations between some processes and quantities that affect galactic evolution or that may be observed as constraints on evolutionary models. The arrows indicate the following process: starting from protogalactic gas, stars are born, evolve and die; they can be observed individually if they are nearby or, otherwise, in integrated light. At their death stars release gas with a changed composition to the interstellar medium (ISM) and perhaps produce cosmic rays, which give rise to interstellar nucleosynthesis. The mass and physical state of the interstellar medium affect the stellar birthrate as well as the observed colors and luminosities (by gaseous emission and extinction by dust). Gas flows and stellar motions to and from the system under study are not indicated, but may be very important.

for instance) suffer larger abundance variations than those which are produced in more violent events (explosive nucleosynthesis). This feature will be explained below.

For some elements (mainly C, N, O, S) isotopic ratios can be estimated not only in the Solar System material but also in the interstellar matter of the solar neighborhood and in the Galactic Center. Table IX.1 displays the variations in C, N, O, S isotopes which have been observed up to now (mainly based on molecular measurements made by Penzias *et al.* at Bell Laboratories, Murray Hill, New Jersey). The isotopes ^{13}C, ^{14}N, and ^{17}O seem to be enriched with time with respect to ^{12}C, ^{15}N, and ^{18}O. This effect is more pronounced in the Galactic Center than in the solar neighborhood. The abundances of heavy elements vary not only with time but also with the region which is observed:

(1) In nearby spiral galaxies which are similar to our Galaxy, such as Andromedra (M31), there exists strong abundance gradients for O, S, and especially N. O and S can be enriched by factors up to 10 in inner regions of the disks of spiral galaxies with respect to the outermost regions. This enrichment can be as large as 30 to 50 for N. This effect can be correlated with many factors but one notes that in the central regions of spiral galaxies the ratio μ between the mass of gas and the total mass (per unit volume) decreases (the mass in form of stars is more important). For instance, μ goes from 0.10 in the solar neighborhood to <0.01 within 300 pc of the galactic center. The decrease of the element abundances with the galactocentric distance is then related to the increase of μ. At this point, one should note that the galaxies

TABLE IX.1

1. *Fractionation*				
Molecules	H_2CO	CO	CS	HCN
$^{12}C/^{13}C$	46 ± 16	46 ± 20	51 ± 13	35 ± 26
Sources*	10	15	5	6

2. *Isotopic ratios*				
Molecular ratios	$^{12}C/^{13}C$	$^{12}C^{18}O/^{13}C^{16}O$	$^{12}C^{17}O/^{12}C^{18}O$	$^{12}C^{15}N/^{13}C^{14}N$
Solar System	89	0.178	0.185	0.32
Solar neighborhood	45	0.074	0.22	0.19
Galactic center	25	0.04	0.31	<0.026

* Number of sources (clouds) in the solar neighborhood.

which have the largest μ (e.g. the case of the Magellanic Clouds and the blue irregular galaxies) are also deficient in heavy elements.

To summarize the observations, there are noticeable variations of the abundances with place and time: places where gas is predominant show lower metal abundances than those where stars dominate. The increase of the abundances with time is more important within the first 10^9 yr of the galactic evolution. This effect is also more important for elements like ^{14}N, ^{13}C or the s process elements which appear to be readily formed by the 'quiet' nucleosynthetic processes.

IX.2. The Ingredients of the Galactic Chemical Evolution

The chemical evolution of a galaxy is mainly governed by the transformation of its gas into stars and by the way those stars evolve and reject part of the gas they processed. Figure IX.2 displays a schematic representation of the processes which can affect the galactic evolution. The important physical parameters therefore are: (1) those which are related to the star formation, their evolution and their nucleosynthetic properties, and (2) those which are related to dynamical effects such as possibilities of inflow of external gas or inhomogeneities of the regions under consideration.

IX.2.1. PARAMETERS RELATED TO THE STELLAR POPULATIONS

Star formation is indeed the governing process of the galactic evolution. The birthrate of a star at a given time t is also a gunction of its mass. This function $\Psi (m, t)$, which is the number of stars of mass m born at time t, is generally separated into a function of mass (hereafter referred to as the initial mass function IMF) and a function of time, although this appears to be a rather crude procedure and is even often wrong

$$\Psi (m, t) = \phi (m) g (t).$$

The separation between the variation with mass and the variation with time is rather artificial: it supposes that the distribution with mass of the stars formed at any time t remains the same throughout the galactic evolution which is not always true. Nevertheless, the advantage of this hypothesis comes from its simplicity.

The function $\phi(m)$, i.e. the initial mass function (IMF), defines the relative proportion of stars of different masses m born at the same time. The IMF is generally assumed to follow a power law of the mass. According to the Salpeter law

$$\phi(m) = \zeta(x - 1) m^{-x},$$

where the parameter x $(1.3 < x < 1.8)$ is derived from the overall luminosity of the stars. ζ is a normalization parameter which expresses the fact that only 25% of the stars are born with a mass $> 1 M_\odot$. The Salpeter law simply describes the fact that low mass stars are more often formed than high mass stars.

The time dependence of the stellar birthrate is related to the time dependence of the gas density. It is quite obvious that if there is no addition of fresh gas during the galactic evolution the gas density is a decreasing function of the time. This is due to the fact that a large fraction of the stellar material does not return to the interstellar gas at the end of the evolution of the star: this is the case of the matter which forms white dwarfs, neutron stars or possibly black holes (Chapter II). μ is the gas density per unit volume such that

$$\alpha S + \mu = 1, \tag{IX.1}$$

where S is the density of matter in the stellar form and α is the proportion of mass in each generation of stars that remains in stellar form ($\beta = 1 - \alpha$ is the proportion of mass which was inside of the stars and is subsequently ejected). The stellar birthrate S is then assumed to be related to the gas density μ according to the Schmidt law:

$$\frac{dS}{dt} \propto \mu \tag{IX.2}$$

with $1 < n < 2$. This expresses simply the fact that there are more stars formed when there is more gas available. Recent observational work based on the distribution of young objects in the Galaxy gives support to this Schmidt law. The combination and integration of Equations (IX.1) and (IX.2) give the evolution of the gas density with respect to the time.

If $\quad n = 1, \quad \mu(t) \sim e^{-t/\tau_0}.$

If $\quad n = 2, \quad \mu(t) \sim \dfrac{1}{1 + t/\tau_0}$ $\tag{IX.3}$

In these two expressions τ_0 is the time scale of matter processing into a star. For example, this time scale is $\sim$ a few 10^9 yr in the solar neighborhood. This time scale is much smaller ($\tau \lesssim$ a few 10^8 yr) in highly processed regions such as the Galactic Center.

The evolution of a given star is directly related to its mass (Chapter II); its lifetime $\tau(m)$ varies roughly like $1/m^3$ where m is the mass of the star.

(ii) Low mass stars, with $m < 3$ to $5M_\odot$ evolve through the planetary nebulae phase to white dwarfs, while at the end of their evolution high mass stars explode as supernovae and leave a remnant which is a neutron star or becomes a black hole. Table IX.2 provides some indication on the rate of stellar deaths according to the mass of the star.

The nucleosynthetic power of a given star, namely its ability to enrich the interstellar gas with heavy elements, depends also on the mass of the star. Figure IX.3 provides an illustration of the mass fractions of different elements which are eventually rejected by the star either during its evolution (stellar evolution) or at the end of it (planetary nebula-supernova phase). For instance, elements with $A \geq 12$ are mainly rejected by stars with $m > 10M_\odot$. At this point, one should make two remarks:

(i) Stars are not the only nucleosynthetic agents: D, ^{3}He, and ^{4}He are presumably formed during the early phase of the Universe while the light elements Li, Be, and B are produced in the interstellar gas by its interaction with cosmic rays.

(ii) As said before, there are elements which are produced in any stars (low and high mass stars) by transformation of nuclei produced in previous stellar generations: for instance, this is the case of ^{13}C and ^{14}N produced from ^{12}C and ^{16}O seeds via the CNO cycle occurring in the H burning zones of the stars. This is also the case of the s process elements produced from Fe in the He burning zone of red giants. These elements which can be found enriched in the outer layers of some planetary nebulae or in some red giants are called secondary elements to distinguish them from primary elements which can be produced in first generation stars.

TABLE IX.2

Stellar deaths (approximate estimates for single stars)

Mass range ($M_\odot$)	Mass loss process	Remnant	Rate in solar neighborhood ($pc^{-2} yr^{-1}$)
$m \lesssim 1$	(Do not die in galactic lifetime)		
1 to m_w ($3 \lesssim m_w \lesssim 8$)	Red giant winds Planetary nebulae	White dwarf	$(1 \text{ to } 4) \times 10^{-9}$
m_w to m_c ($4 < m_c < 10$)	'C detonation supernova'	Neutron star or nothing	Unknown. None if $m_w > m_c$
$m > m_s$ ($m_s = \max[m_w, m_c]$)	Stellar winds Type II supernovae	Neutron star (or black hole?)	$(0.5 \text{ to } 5) \times 10^{-11}$

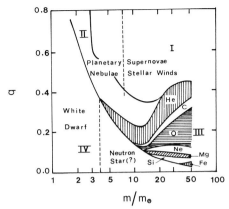

Fig. IX.3.   Illustrative and highly uncertain: stellar mass fractions containing different elements at the time the mass is returned, via planetary nebulae, stellar winds, or supernovae, to the interstellar medium. Zones I and II retain most of the original stellar composition, except for some destruction of light elements (including complete conversion of D to ^{3}He in both zones, and ^{3}He to ^{4}He in zone I), conversion of ^{12}C and ^{16}O to ^{14}N, and enrichment in ^{13}C. In zone III, some He produced during H burning is ejected without further processing, as are heavier elements from later stages of nucleosynthesis. 'Si' refers to elements from Si to Ca, and 'Fe' refers to the iron peak elements. The material in zone IV is locked into remnants.

IX.2.2. DYNAMICAL EFFECTS: INFLOW OF EXTERNAL GAS; INHOMOGENEITIES

It is quite obvious that the dynamical evolution of the galaxies rules their chemical evolution: As we will review later on, it is well known that elliptical galaxies are generally more massive, have a lower gas density and a higher mass-luminosity (M/L) ratio than spiral and irregular ones. Dynamics govern the collapse of the galactic gas and consequently the star formation rate e.g. in spiral galaxies, stars are more likely to be formed in the spiral arms than outside these arms. One should also notice that population I stars (disk population) and population II stars (halo population) are distinct classes not only because of their abundances (lower in population II than in population I) but also evidently because of their dynamical properties, namely their velocity dispersions which are high for population II and low for population I.

Some of the models of chemical evolution make explicit use of effects related to dynamical processes: this is the case (i) for inflows of external gas, and (ii) for possible inhomogeneities of the interstellar gas.

(1) The galaxies cannot always be considered as closed boxes without any relation to their surroundings; our Galaxy may be accreting external gas at a rate of a few per cent of its mass per 10^9 yr. In support of this hypothesis, radioastronomers have observed clouds of neutral H with a high velocity around the galactic disk which seem to fall down this disk. However, this evidence is still controversial. Other observations exist such as the velocities of the interstellar gas observed in front of distant stars which may constitute indirect arguments in favor of this accretion hypothesis. Since this accretion rate is comparable to the rate of star formation, this process, if it takes place, may largely influence the chemical evolution of our Galaxy. We are not sure that this process does actually take place and we do

not know what is the composition of this gas: it could be either unprocessed material with no heavy elements or gas coming from a halo contaminated by population II stars. On the other hand, the discovery of the Fe line in the X ray spectrum emitted by the intergalactic gas of some clusters of galaxies provide evidence that gas can be lost by galaxies: galactic winds may have also important effects on the evolution of galaxies: if the gas disappears the rate of star formation and consequently the rate of chemical contamination may decrease.

(2) The homogeneity on a small scale of the interstellar gas is not fully established. There is a possibility that supernova ejecta do not fully mix with the neighboring material. For instance, there is evidence that isotopic ratios of elements such as O, Ne or Mg are not homogenized in a given class (type II carbonaceous chondrites) of meteorites. As it will be seen later on these effects are also important in the chemical evolution of the galaxies.

IX.3. Evolution of the Solar Neighborhood

IX.3.1. THE SIMPLE MODEL

The Sun is rotating around the Galactic Center with a period of $\sim 10^8$ yr. Therefore, we define the solar neighborhood as "the torus region around the Galactic Center generated by the rotation of the region within 1 kpc from the Sun." In this region, the interstellar gas is assumed to have a homogeneous composition. To study the evolution of this region, a simple model has been designed in which the evolution can be analytically followed.

In the simple models the following assumptions are made:

(1) At the beginning of the evolution of the solar neighborhood ($t = 0$), one assumes that all the galactic material was gaseous and without any heavy elements (here heavy elements are those with atomic mass $A \geq 12$).

In other words at $t = 0$, $\mu(0) = 1$, $S(0) = Z(0) = 0$, where μ, S, and Z are respectively the mass fraction in the form of gas, the mass fraction in the form of stars and the mass fraction of heavy elements.

At the present time, $t_1 \sim 12 \times 10^9$ yr, $\mu_1 = 0.10$, $Z_1 = 0.02$.

(2) The rate of star formation is assumed to follow the Schmidt law.

(3) One also assumes that the initial mass function (IMF) is time independent and one adopts the Salpeter law for it.

(4) As can be seen from Figure IX.4, heavy elements $A \geq 12$ are, for the majority, formed by high mass stars which have a short lifetime (the exception are the secondary elements ^{14}N, ^{13}C, s elements, for which low mass stars are important regarding their evolution). When one analyzes the enrichment into heavy elements one can neglect the lifetime of the stars responsible for the nucleosynthesis. This assumption is called *the instant recycling approximation*.

In the frame of this last assumption and *only in this case*, one can define a very useful parameter called the yield p: this yield represents the mass of metals which is rejected into the interstellar medium per unit mass of gas which is *locked* into stars. If α is the fraction of matter forming the stars and $\beta = 1 - \alpha$ the fraction of interstellar gas which can be rejected and enriched into heavy elements, the yield

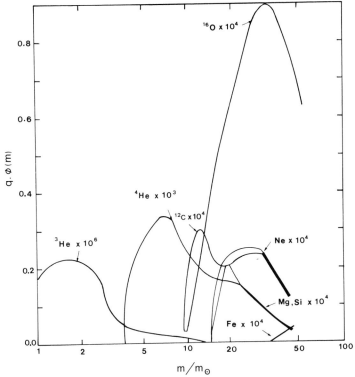

Fig. IX.4.   Relative contributions to enrichment in various elements from stars of different masses; this figure is derived from Figure IX.3 using an initial mass function (IMF)

$$m\left[\frac{dN}{dm}(\text{stars})\right] = 0.25(x-1)\,m^{-x}(m > 1M_\odot) \quad \text{with} \quad x = 1.55.$$

The quantities illustrated here should be considered as highly schematic.

p is $p = p'/\alpha$, where p' is the mass fraction of heavy elements in the interstellar gas. In the simple model describing the solar neighborhood $\alpha \sim 0.8$ and $p = 0.01$. As is shown here the yield is in fact the best parameter to evaluate the evolution of the heavy element abundances.

The simple model also assumes that the region for which the evolution is studied, is isolated. For the moment, one does not consider any inflow of external gas. Under these conditions, Equation (IX.1) can be used again:

$$\mu = 1 - \alpha S.$$

If Equation (IX.1) is combined with the Schmidt law, one obtains:

$$\frac{d}{dt}\ln\left(\frac{1}{\mu}\right) = \frac{1}{\tau} = \mu\frac{\mu^{n-1}}{\tau_0}, \tag{IX.4}$$

where τ_0 is the characteristic time of transformation of gas into stars. For example, as shown by Pagel and Patchett (1975), if $n = 1$ the characteristic time is:

$$\tau = \tau_0 = \frac{t_1}{\ln(1/\mu_1)} \sim 5 \times 10^9 \text{ yr}$$

and the present rate of star formation:

$$\frac{dS}{dt_1} \sim \frac{S}{t_1} \frac{\mu_1}{1 - \mu_1} \ln\left(\frac{1}{\mu_1}\right) \sim 0.25 \frac{S_1}{t_1} : 3M_\odot/\text{yr} \qquad \text{(IX.5)}$$

for the whole Galaxy.

The evolution with time of the abundance of the heavy elements is given by:

$$\frac{d(Z\mu)}{dt} = p' + Z(\beta - p') - Z. \qquad \text{(IX.6)}$$

In this equation, the first term represents the enrichment into heavy elements due to the gas processing stars; the second term, the amount of heavy elements which have been processed into stars but have been rejected by the gas returning to the interstellar medium; the last term is the amount of heavy elements which went from the gas into the stars.

Equation (IX.6) can be easily transformed into:

$$\frac{dZ}{d\ln(1/\mu)} = p(1 - Z) \cong p. \qquad \text{(IX.7)}$$

From this expression one can see how useful the definition of the yield is and also that the abundance of the heavy elements significantly increases with t because of the decrease of the gas density. In fact, in this result lies one of the main difficulties of the simple model: actually, the variation of the observed abundances of the heavy elements is not significant for times $> 10^9$ yr.

Another observable quantity is the number of stars $S(Z)$ which have at their surface an abundance Z of heavy elements. Pagel and Patchett (1975) showed that the ratio of the number of stars of metallicity Z and of present metallicity Z_p is:

$$\frac{S(Z)}{S_1} = \frac{1 - \mu_1}{1 - \mu_1}$$

In Figure IX.5 the ratio Z/Z_1 is plotted against the ratio $S(Z/Z_1)$: the calculated curve is in strong disagreement with the observations because: (i) it does not meet the observed points, and (ii) its curvature is opposite to the one of a curve crossing these points. In other words, the calculations predict too large a number of stars with low metallicity. Note also that not a single star has been yet observed with no heavy elements at all ($Z = 0$). This discrepancy between the evolution of the metallicity of stars, observed and predicted by the simple model, remains the second difficulty of this model.

To solve these two problems (the level off of the gas metallicity and the distribution of stars according to their metallicity) different models have been proposed in which some of the previous hypotheses have been abandoned.

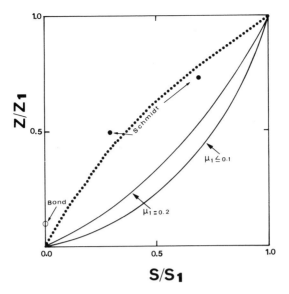

Fig. IX.5. Cumulative distribution of heavy element abundance among long lived stars according to the simple model (full lines) where μ is the present gas mass fraction, and one variable initial mass function model (dotted line). Observational points are from Schmidt (1963) and Bond (1970).

IX.3.2. FURTHER MODELS OF EVOLUTION

IX.3.2.1. *Variable Initial Mass Function Models (VIMF)*

In the simple model it is assumed that the distribution of the stars at their birth according to their initial mass function (IMF) is constant all throughout the galactic evolution. In the VIMF model, the distribution (IMF) is supposed to vary with time: for instance, one can assume that it is more and more difficult for massive stars to be formed when the Galaxy becomes older. The result of such a modification in the galactic model is that it can reproduce the saturation of the gas metallicity as well as the small number of low Z stars. This is because the production of heavy elements, which is governed by the high mass stars, would mainly occur at the beginning of the evolution. One can notice in Figure IX.6 that the curve calculated from the VIMF model has a curvature which agrees with the observed curvature of the $Z - S(Z)$ distribution.

IX.3.2.2. *Prompt Initial Enrichment (PIE Model)*

In these models, it is assumed that a fair amount of heavy elements is produced at the beginning of the galactic evolution. This can be obtained if an outburst of massive stars occurs before/or during the formation of the Galaxy. It would explain the distribution of the gaseous and stellar metallicities and also the fact that there is no star with $Z = 0$.

IX.3.2.3. *Metal Enhanced Star Formation (MESF)*

In this case, it is assumed that the process of star formation more likely occurs in

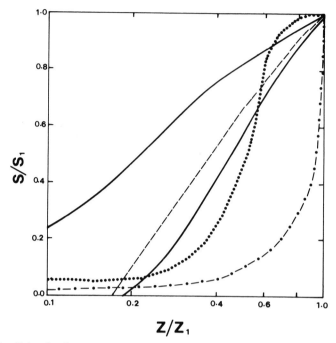

Fig. IX.6.   Metallicity distribution. S/S_1 is the fraction of G and K dwarfs in the solar neighborhood with a metal abundance less than Z, where Z_1 is the present interstellar abundance. *Lower solid line*: schematic representation of the data after removing an estimated dispersion due to observational errors (after Pagel and Patchett, 1975). *Upper solid line*: the 'simple model'. *Dashed line*: effect of a finite initial abundance, $Z_0 = 0.17Z$ (PIE). *Dash-dotted line*: an infall model. *Dotted line*: the infall model with a gaussian distribution of Z at all times, with $\sigma(\log Z) = 0.2$. In cases, Z_1 is the value at which $S/S_1 \approx 1$.

regions where the metal abundance is locally higher (the heavy elements would increase the opacity of the gas then make its cooling and therefore its collapse more easy). With these models the metallicity distribution is also reproduced in a better way.

IX.3.2.4. *Accretion or Infall of External Gas*

In these models, the solar neighborhood is no more a closed box without external interaction. Actually, this region can receive either external intergalactic unprocessed gas, (which therefore does not have any heavy elements), or some gas coming from other parts of the Galaxy such as a halo, in this case the accreted gas can be enriched in metals. Infall models of intergalactic gas reproduce rather well the plateau of the metal distribution in the gas and the small number of low Z stars but cannot explain the curvature of the $S - S(Z)$ distribution.

To summarize this short review of the analytical models, let us say that the first difficulty of the Simple Model namely the level off of Z with t, can be relaxed either by the VIMF model or better by the PIE model (birth and death of massive stars at the very beginning) or by the infall model. An alternative model is to assume that massive stars can 'swallow' the nucleosynthetic products either by becoming black holes or by favoring the formation of solid bodies and making the largest part of the metals invisible.

As for the second difficulty (the small number of low Z stars), the best approaches are the PIE or MESF models.

IX.4. Numerical Models With No Instant Recycling

All of the previously discussed models are developed in the frame of the instant recycling approximation which assumes that the stars have a negligible lifetime compared to the evolution time scale of the Galaxy. Actually, such an approximation does not hold in the two following cases where the presence of low mass stars is important for the evolution:

(1) As said before there are some elements, especially secondary elements, produced in stars of subsequent generations (secondary ...) which are largely enriched in the external zones of low mass stars: this is the case of ^{13}C, ^{14}N, and of the s process elements for instance.

(2) The instant recycling approximation does not apply to highly processed re-regions where the gas density becomes very small. In that case, the gas density in later phases is governed by the rate of mass ejection suffered by low mass stars (long lived).

Under such circumstances, one is led to develop numerical models where the stellar lifetimes are properly taken into account.

These models have been used in attempting to reproduce the evolution of the gas density, the rate of mass ejected by stars and the evolution of the CNO isotope abundances, not only in the solar neighborhood but also in the Galactic Center which is a highly processed region (low gas density) where low mass stars are more numerous than in the solar neighborhood (see Table IX.3). Although the Galactic Center is optically invisible, the gas density can be estimated by comparison with the CO abundance; the rate of death of high mass stars can be approached by counting the number of large nonthermal radio sources which are assumed to be supernova remnants. Finally, the rate of planetary nebulae can be evaluated by analyzing the pattern of IR sources.

TABLE IX.3

Compared properties of the Galactic Center (inside ~ 200 pc) and the solar neighborhood. Supernovae explosions monitor the rate of formation and death of massive stars ($M \gtrsim 5M_\odot$), planetary nebulae monitor the death rate of less massive stars ($1M_\odot < M \lesssim 5M_\odot$).

	Galactic center	Solar neighborhood
Gas/star mass ratio	$(2 \text{ to } 10) \times 10^{-3}$	$(1 \text{ to } 1.5) \times 10^{-1}$
Rate of supernovae explosions		
— per unit total mass	$(2 \text{ to } 15) \times 10^{-14} \text{ yr}^{-1} M_\odot^{-1}$	$(2 \text{ to } 8) \times 10^{-13} \text{ yr}^{-1} M_\odot^{-1}$
—per unit mass of gas	$(0.2 \text{ to } 8) \times 10^{-10} \text{ yr}^{-1} M_\odot^{-1}$	$(1.3 \text{ to } 8) \times 10^{-12} \text{ yr } M_\odot^{-1}$
Rate of formation of planetary nebulae		
— per unit total mass	$(0.7 \text{ to } 6) \times 10^{-11} \text{ yr}^{-1} M_\odot^{-1}$	$(0.5 \text{ to } 2) \times 10^{-11} \text{ yr}^{-1} M_\odot^{-1}$
—per unit mass of gas	$(0.7 \text{ to } 30) \, 10^{-8} \text{ yr}^{-1} M_\odot^{-1}$	$(0.3 \text{ to } 2) \times 10^{-10} \text{ yr}^{-1} M_\odot^{-1}$

Figure IX.7 shows the evolution of the gas density calculated to mimic the observa-
in the solar neighborhood and in the Galactic Center. One can easily see the importance of the mass ejected by the low mass stars in the evolution of the gas after a few
10^9 yr: the slope of the curve $\mu(t)$ becomes less steep than in the case of the instant
recycling approximation because of the importance of the mass released by these long
lived stars. It is also evident that low mass stars are the main contributors to the gas
reprocessing within the interstellar medium. At the beginning of the galactic evolution, the gas is mainly rejected by the high mass short lived stars. It takes 2×10^9 yr
for the low mass stars to release more gas than the high mass stars in the solar neighborhood and only 2×10^8 yr in the Galactic Center (Figure IX.8).

It is not the purpose of this chapter to describe in too much detail the evolution
of the isotopic ratios such as $^{12}C/^{13}C$ and $^{15}N/^{14}N$ in the interstellar medium. The
only comment which can be made in this respect is that the numerical models described above (Vigroux *et al.*, 1976) reproduce well the observed $^{12}C/^{13}C$ and
$^{15}N/^{14}N$ ratios. In particular, ^{13}C is enriched by about a factor 2 relative to ^{12}C
during the last 5×10^9 yr in the solar neighborhood. In the Galactic Center, this
enrichment can be about a factor of 4. Furthermore, this model also explains that
ratio N/O is 3 to 4 times larger in the Galactic Center than in the solar neighborhood.

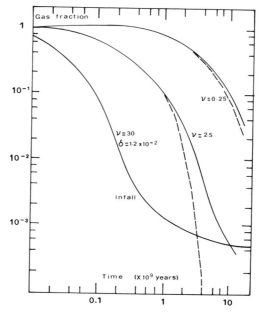

Fig. IX.7. Evolution of the gas fraction σ for the solar neighborhood ($v \sim 0.2$ to 0.5) where $\sigma \sim 0.1$
at $t \sim 10^{10}$ yr and the Galactic Center ($v > 2$) where $\sigma \sim 5 \times 10^{-4}$ at $t \sim 10^{10}$ yr. Dashed lines represent
the gas evolution when calculations are made with instant recycling: notice the rather important discrep-
ancy when $v > 2$. The curve labeled infall shows the gas evolution for $v = 30$ and a rate of infall of $\sim 2M_\odot$
of gas per year for the whole Galaxy. Infalling material sets the limit for σ at large times.

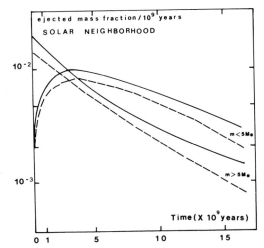

Fig. IX.8a.   Mass fractions ejected from low mass stars ($M < 5M_\odot$) and high mass stars ($M > 5M_\odot$) in the solar neighborhood. Calculations have been made for a model with $v = 0.25$ and the same rate of infall as in Figure IX.7 (solid lines) and for a model with $v = 0.2$ and no infall (dashed lines). Note that in both cases the mass ejected by low mass stars becomes larger than that ejected by high mass stars for $t \sim 2 \times 10^9$ yr.

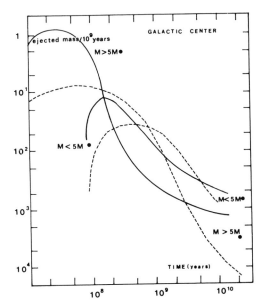

Fig. IX.8b.   Mass fractions ejected from low mass stars ($M < 5M_\odot$) and high mass stars ($M > 5M_\odot$) in the Galactic Center. *Dashed line*: $v = 2.5$, no infall; *solid line*: $v = 30$ and the same rate of infall as in Figure IX.7. Notice that infall and large values of v do not affect significantly the mass fraction ejected by low mass stars but modify the mass fraction ejected by large mass stars proportionally to v.

Finally, the ratio $^{12}C\ ^{15}N/^{13}C\ ^{14}N$ decreases more than a factor 10 larger in the solar vicinity than in the Galactic Center, provided that ^{15}N is produced in massive stars. All these features are consistent with the fact that low mass stars producing ^{13}C and ^{14}N are relatively more numerous in the central regions than in the solar neighborhood. They assume that the rate of infall, if it does exist, is moderate: $\delta < 10^{-3}M_\odot/M_\odot$ (total) per 10^9 yr (i.e. $<0.1M_\odot/$yr for the whole Galaxy).

Current models of the Galactic Center are also able to account for the D abundance in this region: $D/H_{\text{gal. cent.}} \sim 0.1 \times (D/H)_{\text{solar neighb.}}$ by adopting at least one of the two following assumptions:

(i) The Galactic Cosmic Rays (GCR) have a larger flux in the central regions than in the solar neighborhood, a richer spectrum in particles of rather low energy (a few tens MeV). In this case an abundance $D/H \sim 10^{-6}$ can easily be accounted for by these fluxes. There is indeed indirect information that the GCR flux can be larger because the high energy γ ray flux ($E \geq 70$ MeV) is somewhat larger in the central regions than in the solar neighborhood. The only way to test the presence of GCR particles of a few tens MeV is to look for γ ray lines coming from nuclear reactions induced by these particles. In the present situation, this hypothesis appears to be quite reasonable.

(ii) If there are not enough GCR particles to produce D spallation reactions in the Galactic Center, it must be assumed that the Galactic Center experiences some inflow of external gas. If there is no infall, the D abundance is given by $d\sigma_D/dt = -v\sigma_D$ and with values of the parameter v adapted to fit the gas content of the central regions, the D/H ratio would be as low as 10^{-12} for the regions considered. The adjunction of external gas with primordial D abundance, at a very modest rate $\delta \leq 10^{-3}\ M_\odot/M_\odot$ of total mass (in 10^9 yr), is also able to explain the observed abundance of D in the Galactic Center.

To conclude this section devoted to the review of the results of the numerical models, let us say that although these models are slightly more difficult to work out, they provide the only way to study the evolution of regions like the Galactic Center. Furthermore, to our knowledge, there is no other way to describe the evolution of secondary elements produced by low mass stars, such as ^{13}C or ^{14}N. In the case of the Galactic Center, it is fair to remark that such models are certainly far too crude to provide a correct account of the evolution of such highly processed regions. Dynamical analyses, i.e., not only an account of possible infall of external material but also a hydrodynamical treatment of the physics of the gas and of the stellar component, should be undertaken to elaborate a better modelization of these regions.

IX.5. Conclusion: Chemical Evolution of Galaxies

We have concentrated our review on the evolution of our own Galaxy, since it is the one we know the best. Nevertheless, a few remarks can be made concerning the evolution of the other galaxies.

(i) Gradients in the abundances of the heavy elements have been seen in a number of nearby spiral galaxies and can be explained by models similar to those described in the previous section.

(ii) Concerning their chemical evolution, the elliptical galaxies may be reproduced by models similar to those which mimic the evolution of the Galactic Center of galaxies.

(iii) Recent X-ray observations performed for different clusters (Coma, Virgo, . . .) show the presence of Fe in abundances ~ 0.3 to 0.4 of the 'standard abundance' in the *intergalactic gas*. This can be explained by the sweeping of the galactic gas out of some galaxies which have been processed and enriched into heavy elements at the very early evolution of the galactic clusters. The discovery of '*anemic galaxies*' with low gas densities is also a proof of the existence of galactic winds which could affect the chemical evolution of galaxies.

(iv) The Magellanic Clouds which are satellites of our own Galaxy are known to have larger gas contents and lower metallicities than the Galaxy itself. The heavy elements have abundances 3 to 5 times lower in the Large Magellanic Cloud and about 10 times lower in the Small Magellanic Cloud than in our own Galaxy. Meanwhile, He seems to have a slightly lower abundance corresponding to a depletion of He (with respect to our Galaxy) $\Delta Y \sim 3 \Delta Z$ which is predicted from the usual models presented here. Furthermore, the N/O ratio seems to be the same as the one found in the solar neighborhood. This would mean that the processes of star formation may have taken place there more recently than in our Galaxy. More generally, the existence of blue compact galaxies is explained in terms of sporadic processes of star formation: the color of a galaxy depends on the stellar populations, for instance, the blue color is the signature of places where young massive stars are formed.

The main interest of the studies of galactic evolution lies in the fact that one has to integrate patchy and apparently unrelated observations to attempt to understand such an evolution. One should be aware that the number of uncertain results such as crude observations or ill-defined estimates still prevents us from giving a clear cut view of such evolution processes. However, the development of new observational techniques, like the Space Telescope, and hopefully a better understanding of the physics and chemistry of the interstellar medium and of star formation, should allow significant progress in this exciting field which needs the whole knowledge of astrophysics to be worked out satisfactorily.

References

Quoted in the text:

Audouze, J. and Tinsley, B. M.: 1976, *Ann. Rev. Astron. Astrophys.* **14**, 43.
Audouze, J., Lequeux, J., and Vigroux, L.: 1975, *Astron. Astrophys.* **43**, 71.
Pagel, B. E. J. and Patchett, B. E., 1975, *Monthly Notices Roy. Astron. Soc.* **172**, 13.
Vigroux, L., Audouze, J., and Lequeux, J.: 1976, *Astron. Astrophys.* **52**, 1.
Wannier, P. G.: 1977, in J. Audouze (ed.), *CNO Isotopes in Astrophysics*, D. Reidel Publ. Co., Dordrecht, Holland.

Further readings:

Audouze, J., Lequeux, J., Reeves, H., and Vigroux, L., 1976, *Astrophys. J.* **208**, L51–L54.

Audouze, J. and Tinsley, B. M.: 1974, *Astrophys. J.* **192**, 487.

Basinska Greszik, E. and Mayor, M., 1978, 'Chemical and Dynamical Evolution of our Galaxy', *IAU Colloq.* **45**, Torun, Poland, Sept. 1977, (Sauverny Observatoire de Genève press).

Guibert, J., Lequeux, J., and Viallefond, F.: 1978, *Astron. Astrophys.* **68**, 1.

Ostriker, J. P. and Thuan, T. X.: 1975, *Astrophys. J.* **202**, 353.

Peimbert, M.: 1975, *Ann. Rev. Astron. Astrophys.* **13**, 113.

Talbot, R. J. and Arnett, W. D.: 1971, *Astrophys. J.* **170**, 409.

Talbot, R. J. and Arnett, W. D.: 1973b, *Astrophys. J.* **186**, 69.

Tinsley, B. M.: 1972, *Astron. Astrophys.* **20**, 383.

Tinsley, B. M. and Larson, R. B. (ed.): 1977, *The Evolution of Galaxies and Stellar Populations*, Yale University Observatory, New Haven, U.S.A.

Trimble, V.: 1975, *Rev. Mod. Phys.* **47**, 877.

Truran, J. W. and Cameron, A. G. W.: 1971, *Astrophys. Space Sci.* **14**, 179.

Vigroux, L.: 1977, *Astrophys. Letters* **56**, 473.

CONCLUSION

Modern astrophysics made tremendous progress around 1938, when Bethe and Von Weiszäcker discovered that under specific conditions four H nuclei can be transmuted into one He nucleus. The energy released by this fusion reaction is large enough to account for the luminosity of the Sun and similar stars. There is enough H fuel inside the Sun to allow it to last 10^{10} yr with its observed luminosity. The age of the Sun (4.6×10^9 yr), which is a consequence of this nuclear process, is in accordance with the work of Rutherford who analyzed Pb isotopes in rocks early in this century. He was the first to demonstrate that the Earth and consequently the whole Solar System has an age of several billion years.

The discovery of the H burning cycle in stars was the starting point of nuclear astrophysics. The historical development of nuclear astrophysics can be roughly separated into four different eras:

(a) *The Gamow era* lasted up to the early fifties when all chemical species were assumed to be synthetized during the primordial phases of the Universe. In the earliest versions of these theories, matter was assumed to be born as neutrons which could coalesce and generate all the observed nuclear species. This theory was found to be wrong because: (i) it cannot explain the observed abundance distribution of the elements; and (ii) we know from physics of elementary particles that such a neutron sea cannot survive long enough to be able to synthetize heavy elements.

In a more correct version of the Gamow theory, worked out in particular by Wagoner, the nucleosynthesis which begins with an interaction between protons and neutrons stops at ^{4}He (and possibly ^{7}Li) because of the two short lifetimes of the nuclei with atomic masses of 5 and 8.

Nevertheless, as shown in Chapter VII, the nucleosynthesis occurring in primordial phases is at present the best theory to account for the formation of the lightest elements, H, D, ^{3}He, ^{4}He and very likely ^{7}Li.

(b) *The stellar era* culminated with the publication of the famous Burbidge *et al.* (1957) and the Cameron (1957) lecture notes. One could say that this nucleosynthesis theory, which explains the synthesis of the whole chemical chart in terms of stellar processes has made the field quite fashionable. Historically, this 'stellar' nucleosynthesis theory has been inspired from the steady state cosmological model devoloped by Bondi, Gold and Hoyle: they supposed that the Universe had not experienced any primordial phase: matter was continuously created. This model was fashionable until the discovery of the 2.7 K black body radiation by Penzias and Wilson in 1965 for which they received the Nobel Prize 1978.

Since 1957, nuclear astrophysicists continue to use the B^{2}FH classification of nucleosynthetic processes: H and He burning, α process, equilibrium processes, s, p, and r processes. Let us emphasize again the importance in the He burning phase of Hoyle's prediction of the second excited state of ^{12}C, which has been subsequently observed in nuclear physics experiment. Moreover, the determination of T_C in the spectra of some S stars, the abundances of the s process elements (compared to the neutron absorption cross sections), the evolution of s process element abundances in FG Sagittae constitute nondisputable proofs of the existence of nucleosynthetic processes occurring during the stellar evolution.

(c) *The explosive era*: since 1965–1970 the former graduate students of the previously quoted researchers (Arnett, Clayton, Truran and their students) expanded the idea that many nucleosynthetic processes occur not only in quiet phases (the Main Sequence and the Red Giant phases) but also during the short but energetic last phases of stellar evolution. All throughout this monograph, it has been recalled that large mass stars ($M > 5M_\odot$) end up their evolution as explosive supernovae. Around 1970, it was most fashionable to think that a large part of nucleosynthesis triggered and occurred during the explosion of these very energetic objects. These theories (developed in Chapter V) have been quite successful in reproducing the Solar System abundances of many different elements from C up to Fe and also the p and the r elements. The only, but serious, weakness of these models lies in the fact that the physics of the supernovae and the novae explosions is not yet fully understood and that there are still too many free parameters on which one can play to obtain a reasonable set of abundances.

Meanwhile, other energetic processes which involve spallation reactions between Galactic Cosmic Rays and the interstellar medium have been proposed as responsible for the formation of the Li, Be, and B light elements. It has been shown in Chapter VII that the processes successfully explain both this nucleosynthesis and some properties of the Cosmic Rays.

(d) *The present era*: starts around 1973 and is marked by two types of developments:

(i) the integration of the nucleosynthesis into the more general problem of the galactic evolution (Chapter IX). After a description of the general schemes by which chemical elements are formed in the Big Bang, by the cosmic rays or inside the stars, it seems now important to follow the evolution with time of these abundances during the history of the Universe and of the Galaxy. This topic is a fascinating project in the sense that the whole field of astronomy and astrophysics is involved in it. It is obviously related to the origin and the evolution of the Universe, to the dynamic evolution of galaxies, to the formation and the evolution of stars. The observations which challenge the current theories on chemical evolution of galaxies are the distributions of stellar metallicities, the new determinations of isotopic ratios in the nearby interstellar medium and in the galactic center, the discovery of gradients in the element abundance distributions in nearby spirals and our own Galaxy. For example, the fact that no star with zero metallicity has been observed (with the exception of white dwarfs, a very special case) constitutes a real problem for the theoreticians. Did a first generation of very massive and therefore rapidly evolving stars form in the early phases of the Universe before the formation of all the presently observed stars? Did the first generation of stars exist before the formation of galaxies? . . .

(ii) The discovery reported in Chapter VIII of important isotope anomalies in some mineral phases of carbonaceous chondrites has shown that the nucleosynthetic history must be much more complicated than was thought before, in particular for the Solar System itself. Either the formation of the Solar System has been triggered by the explosion of one or more nearby supernovae or there is an inhomogeneous admixture in the Solar System of interstellar grains formed in various nucleosynthetic sites. Although this problem is not yet solved, there is a great hope that these discoveries of isotope anomalies will provide the best clues for understanding the formation of the Solar System. Since the nucleosynthetic theories have been developed to account in particular for the Solar System composition (the p process elements have only been observed there!) one expects that the whole nucleosynthesis is far more complex than the one described in these pages.

Thanks to Nuclear Astrophysics, which requires the active support of both experimental and theoretical nuclear physics, many general astrophysical problems are now close to being understood, namely the origin and the evolution of the elements, and therefore the observed matter. However, some exciting puzzles remain to be solved:

(1) There is obviously a growing need for a better knowledge of many nuclear physics parameters such as the fusion and neutron absorption cross sections and for a better determination of nuclear masses. Many more nuclear reaction cross sections are needed for explosive nucleosynthesis calculations (occurring either during the Big Bang or at the end of the stellar evolution) than for the previous equilibrium calculations. The physics of evolved celestial bodies such as pulsars, supernovae, and white dwarfs will only be understood when equations of state for the nuclear matter are obtained.

(2) The physics of the weak interactions plays a more and more important role in astrophysics and the new advances of elementary particle physics make that subject more and more exciting. The weak interactions govern the outcome of the Big Bang nucleosynthesis, the transport of neutrinos in central regions of supernovae might be one of the basic mechanisms triggering their explosions. More classical is the yet unsolved problem of the solar neutrinos: only one third of the expected neutrino flux has been detected by the Brookhaven experiment (see page 53). This experiment depends entirely on the $^{37}Cl-^{37}A$ reaction which is only able to detect the high energy neutrinos coming from the ^8B decay. As it has been stated in Chapter IV, the production rate of these neutrinos largely depends on the solar models. New experiments, able to detect lower energy neutrinos should be built up in the near future. Unfortunately, these new experiments which are badly needed in order to understand one of the most basic problems of astrophysics will be quite expensive since they require metals as rare as In and Ga.

(3) The problems which have to be solved by nuclear astrophysics also require a better knowledge of various aspects of more classical physics such as thermodynamics and hydrodynamics. Stellar evolution and especially its latest stages (supernova explosions, planetary nebula formation) are obviously governed by hydro-

dynamical processes. A very important problem is the stellar mass loss: it is observationally known that very hot stars eject a wind of particles much stronger than the solar wind. The processes responsible for these stellar winds are not yet completely understood. Also our present knowledge on the formation of stars and galaxies is poor and tremendous progress in that field of physics is needed.

(4) As it has been obvious in several parts of the monograph, chemistry also plays a major role in the physics of the interstellar medium where molecules and dust are the major constituents. Moreover, the formation of the Solar System (where most abundance is determined) is obviously dominated by chemical processes.

The present time is marked by important theoretical progress, by the fact that many large telescopes (CFH, ESO, Kitt Peak, Caucasus) are operating and also by the launching of extremely exciting space missions (Copernicus, International Ultra-Violet Explorer, Space Telescope, Voyager, Spacelab . . .). There are numerous questions which have to be approached by Nuclear Astrophysics, in which theoretical approaches should complete genuine observational discoveries. With the use of new large telescopes, and especially the Space Telescope which will free the astronomers from the terrestrial atmosphere and allow them to detect all kinds of radiations not visible on the ground, we think that we are now entering a new and fascinating era of Astronomy.

APPENDIX A

I. Some Fundamental Constants

Velocity of light	c	$= 2.997\,925 \times 10^{10}$ cm s^{-1}
Gravitation constant	G	$= 6.67 \times 10^{-18}$ dyn cm^2 g^{-2}
Planck constant	h	$= 6.626\,20 \times 10^{-27}$ erg s
Electron charge	e	$= 4.803\,25 \times 10^{-10}$ ESU*
		$= 1.602\,192 \times 10^{-20}$ EMU*
Mass of electron	m_e	$= 9.109\,56 \times 10^{-28}$ g
Rest mass energy of electron	$m_e c^2$	$= 0.511\,004$ MeV
Mass of proton	m_p	$= 1.672\,661 \times 10^{-24}$ g
Rest mass energy of proton	$m_p c^2$	$= 938.285$ MeV
Mass of Unit Atomic Weight (uma)	M_u	$= 1.660\,531 \times 10^{-24}$ g
(^{12}C = 12 scale)		
Mass energy of Unit Atomic Weight	$M_u c^2$	$= 931.481$ MeV
Boltzmann constant	k	$= 1.380\,62 \times 10^{-16}$ erg deg^{-1}
Avogadro number	N_A	$= 6.022\,17 \times 10^{13}$ mole^{-1}

II. Some Astronomical Constants

Mean Sun-Earth distance (Astronomical Unit)	AU	$= 1.495\,979 \times 10^{13}$ cm
Parsec	pc	$= 3.085\,678 \times 10^{18}$ cm
		$= 3.261\,633$ light yr
Light year		$= 9.460\,530 \times 10^{17}$ cm
Solar mass	$M_\odot$	$= 1.989 \times 10^{33}$ g
Solar radius	$R_\odot$	$= 6.9599 \times 10^{10}$ cm
Solar luminosity	$L_\odot$	$= 3.826 \times 10^{33}$ erg
Earth mass	$M_\oplus$	$= 5.976 \times 10^{27}$ g
Earth equatorial radius	$R_\oplus$	$= 6378.164$ km

---

*   ESU = electrostatic units.
  EMU = electromagnetic units.
  1 ESU = 1 EMU × velocity of light.

III. Some Quantities Associated with One Electron-Volt

$$[E_0 = h\nu_0 = hc/\lambda_0 = hck_0 = kT_0]$$

Energy of 1 eV	E_0	$= 1.602\ 192 \times 10^{-12}$ erg
Wavelength associated with 1 eV	λ_{-1}	$= 12398.54 \times 10^{-8}$ cm
Wavenumber associated with 1 eV	k_0	$= 8066.02$ cm^{-1}
Frequency associated with 1 eV	ν_0	$= 2.417965 \times 10^{14}$ s^{-1}
Temperature associated with 1 eV	T_0	$= 11604.8$ K

APPENDIX B

Atomic Mass Excesses

The atomic mass excess in energy units of an atom of charge number Z (= number of protons in the nucleus) and mass number A (= number of protons + number of neutrons in the nucleus) is:

$$\Delta M = (M - AM_u)c^2,$$

where M is the mass of the atom, M_u the mass of the Unit Atomic Weight (= 1/12 of the mass of a ^{12}C atom), and c the velocity of light.

TABLE OF ATOMIC MASS EXCESSES (after Clayton, 1968)

Z	Element	A	ΔM(MeV)	Z	Element	A	ΔM(MeV)
0	n	1	8.071 44			12	13.370 20
1	H	1	7.288 99			13	16.561 60
	D	2	13.135 91	6	C	9	28.990 00
	T	3	14.949 95			10	15.658 00
	H	4	28.220 00			11	10.648 40
		5	31.090 00			12	0
2	He	3	14.931 34			13	3.124 60
		4	2.424 75			14	3.019 82
		5	11.454 00			15	9.873 20
		6	17.598 20	7	N	12	17.364 00
		7	26.030 00			13	5.345 20
		8	32.000 00			14	2.863 73
3	Li	5	11.679 00			15	0.100 40
		6	14.088 40			16	5.685 10
		7	14.907 30			17	7.871 00
		8	20.946 20	8	O	14	8.008 00
		9	24.965 00			15	2.859 90
4	Be	6	18.375 60			16	− 4.736 55
		7	15.768 90			17	− 0.807 70
		8	4.944 20			18	10.782 43
		9	11.350 50			19	3.332 70
		10	12.607 00			20	3.799 00
		11	20.181 00	9	F	16	10.904 00
5	B	7	27.990 00			17	1.951 90
		8	22.923 10			18	0.872 40
		9	12.418 60			19	− 1.486 00
		10	12.052 20			20	− 0.011 90
		11	8.667 68			21	− 0.046 00

Z	Element	A	ΔM(MeV)	Z	Element	A	ΔM(MeV)
10	Ne	18	5.319 30	17	Cl	32	− 12.810 0
		19	1.752 00			33	− 21.014 0
		20	− 7.041 50			34	− 24.451 0
		21	− 5.729 90			35	− 29.014 5
		22	− 8.024 90			36	− 29.519 6
		23	− 5.148 30			37	− 31.764 8
		24	− 5.949 00			38	− 29.803 0
11	Na	20	8.280 00			39	− 29.800 0
		21	− 2.185 00			40	− 27.500 0
		22	− 5.182 20	18	Ar	34	− 18.394 0
		23	− 9.528 30			35	− 23.051 0
		24	− 8.418 40			36	− 30.231 6
		25	− 9.356 00			37	− 30.950 9
		26	− 7.690 00			38	− 34.718 2
12	Mg	22	− 0.140 00			39	− 33.238 0
		23	− 5.472 40			40	− 35.038 3
		24	− 13.933 30			41	− 33.067 4
		25	− 13.190 70			42	− 34.420 0
		26	− 16.214 20	19	K	36	− 16.730 0
		27	− 14.582 60			37	− 24.810 0
		28	− 15.020 00			38	− 28.786 0
13	Al	24	0.100 0			39	− 33.803 3
		25	− 8.931 0			40	− 33.533 3
		26	− 12.210 8			41	− 35.552 4
		27	− 17.196 1			42	− 35.018 0
		28	− 16.855 4			43	− 36.579 0
		29	− 18.218 0			44	− 35.360 0
		30	− 17.150 0			45	− 36.630 0
14	Si	26	− 7.132 0			46	− 35.340 0
		27	− 12.386 0			47	− 36.250 0
		28	− 21.489 9	20	Ca	38	− 21.690 0
		23	− 21.893 6			39	− 27.300 0
		30	− 24.439 4			40	− 34.847 6
		31	− 22.962 0			41	− 35.140 0
		32	− 24.200 0			42	− 38.539 7
15	P	28	− 7.660 0			43	− 38.395 9
		29	− 16.945 0			44	− 41.459 6
		30	− 20.197 0			45	− 40.808 5
		31	− 24.437 6			46	− 43.138 0
		32	− 24.302 7			47	− 42.347 0
		33	− 26.334 6			48	− 44.216 0
		34	− 24.830 0			49	− 41.288 0
16	S	30	− 14.090 0	21	Sc	40	− 20.900 0
		31	− 18.992 0			41	− 28.645 0
		32	− 26.012 7			42	− 32.141 0
		33	− 26.582 6			43	− 36.174 0
		34	− 29.933 5			44	− 37.813 0
		35	− 28.847 1			45	− 41.060 6
		36	− 30.655 0			46	− 41.755 7
		37	− 27.000 0			47	− 44.326 3
		38	− 26.800 0			48	− 44.505 0

Z	Element	A	ΔM (MeV)	Z	Element	A	ΔM (MeV)
21	Sc	49	− 46.549 0	27	Co	56	− 56.031 0
		50	− 44.960 0			57	− 59.338 9
22	Ti	42	− 25.123 0			58	− 59.838 0
		43	− 29.340 0			59	− 62.232 7
		44	− 37.658 0			60	− 61.651 3
		45	− 39.002 0			61	− 62.930 0
		46	− 44.122 6			62	− 61.528 0
		47	− 44.926 6			63	− 61.920 0
		48	− 48.483 1	28	Ni	56	− 53.899 0
		49	− 48.557 7			57	− 56.104 0
		50	− 51.430 7			58	− 60.228 0
		51	− 49.738 0			59	− 61.158 7
		52	− 49.540 0			60	− 64.470 7
23	V	46	− 37.060 0			61	− 64.220 0
		47	− 42.010 0			62	− 66.748 0
		48	− 44 470 0			63	− 65.516 0
		49	− 47.950 2			64	− 67.106 0
		50	− 49.215 8			65	− 65.137 0
		51	− 52.198 9			66	− 66.055 0
		52	− 51.436 0	29	Cu	58	− 51.659 0
		53	− 52.180 0			59	− 56.359 0
		54	− 49.630 0			60	− 58.346 0
24	Cr	48	− 42.813 0			61	− 61.984 0
		49	− 45.390 0			62	− 62.813 0
		50	− 50.249 0			63	− 65.583 1
		51	− 51.447 2			64	− 65.427 6
		52	− 55.410 7			65	− 67.266 0
		53	− 55.280 7			66	− 66.255 0
		54	− 56.930 5			67	− 67.291 0
		55	− 55.113 0			68	− 65.410 0
		56	− 55.290 0	30	Zn	60	− 54.186 0
25	Mn	50	− 42.648 0			61	− 56.580 0
		51	− 48.260 0			62	− 61.123 0
		52	− 50.702 0			63	− 62.217 0
		53	− 54.682 0			64	− 66.000 3
		54	− 55.552 0			65	− 65.917 0
		55	− 57.704 8			66	− 68.881 0
		56	− 56.903 8			67	− 67.863 0
		57	− 57.480 0			68	− 69.994 0
		58	− 55.650 0			69	− 68.425 0
26	Fe	52	− 48.328 0			70	− 69.550 0
		53	− 50.693 0			71	− 67.520 0
		54	− 56.245 5			72	− 68.144 0
		55	− 57.473 5	31	Ga	63	− 56.720 0
		56	− 60.605 4			64	− 58.928 0
		57	− 60.175 5			65	− 62.658 0
		58	− 62.146 5			66	− 63.706 0
		59	− 60.659 9			67	− 66.865 0
		60	− 61.511 0			68	− 67.074 0
		61	− 59.130 0			69	− 69.326 2
27	Co	54	− 47.994 0			70	− 68.897 0
		55	− 54.014 0				

Two examples of computation of the nuclear energy released by reactions between nuclei:

(1) Energy released by the first weak reaction occurring in stars.

$$p + p \approx D + e^+ + v$$

Atomic mass excess of hydrogen	7.28899
Mass of one electron	0.511
Mass excess of one proton	= 6.77799 MeV

Atomic mass excess of deuterium	13.13591
Mass of one electron	0.511
Mass excess of the deuton (deuterium nucleus)	= 12.62491 MeV

Balance:

Left-hand side 2 × 6.77799	= 13.55598
Right-hand side 12.62491 + 0.511	= 13.13591
	0.42007 MeV

Thus 0.42 MeV are liberated in the form of kinetic energy for the positron and the neutrino. The subsequent annihilation of the positron with an electron puts the total energy released up to $(2 \times 0.511) + 0.42 = 1.442$ MeV per reaction.

(2) Energy released by a strong reaction

$$^{17}O + H \qquad ^{14}N + {}^4He$$

	H	7.28899
	^{17}O	− 0.80770
Atomic mass excess of the incoming particles	=	6.48129

	^4He	2.42475
	^{14}N	2.86373
Atomic mass excess of the products	=	5.28848

The nuclear mass excess would be obtained by subtracting the same number of electrons on both sides, so that the energy released in the reaction is simply:

6.48129
− 5.28848
1.19281

Thus 1.19 MeV are liberated by one reaction.

INDEX OF SUBJECTS

GEOPHYSICS AND ASTROPHYSICS MONOGRAPHS

AN INTERNATIONAL SERIES OF FUNDAMENTAL TEXTBOOKS

Editor:

BILLY M. McCORMAC (Lockheed Palo Alto Research Laboratory)

Editorial Board:

R. GRANT ATHAY (High Altitude Observatory, Boulder)
W. S. BROECKER (Lamont-Doherty Geological Observatory, New York)
P. J. COLEMAN, Jr. (University of California, Los Angeles)
G. T. CSANADY (Woods Hole Oceanographic Institution, Mass.)
D. M. HUNTEN (University of Arizona, Tucson)
C. DE JAGER (the Astronomical Institute at Utrecht, Utrecht)
J. KLECZEK (Czechoslovak Academy of Sciences, Ondřejov)
R. LÜST (President Max-Planck-Gesellschaft zur Förderung der Wissenschaften, München)
R. E. MUNN (University of Toronto, Toronto)
Z. ŠVESTKA (The Astronomical Institute at Utrecht, Utrecht)
G. WEILL (Institute d'Astrophysique, Paris)

Date Due